Materials Horizons: From Nature to Nanomaterials

Series Editor

Vijay Kumar Thakur, School of Aerospace, Transport and Manufacturing, Cranfield University, Cranfield, UK

Materials are an indispensable part of human civilization since the inception of life on earth. With the passage of time, innumerable new materials have been explored as well as developed and the search for new innovative materials continues briskly. Keeping in mind the immense perspectives of various classes of materials, this series aims at providing a comprehensive collection of works across the breadth of materials research at cutting-edge interface of materials science with physics, chemistry, biology and engineering.

This series covers a galaxy of materials ranging from natural materials to nanomaterials. Some of the topics include but not limited to: biological materials, biomimetic materials, ceramics, composites, coatings, functional materials, glasses, inorganic materials, inorganic-organic hybrids, metals, membranes, magnetic materials, manufacturing of materials, nanomaterials, organic materials and pigments to name a few. The series provides most timely and comprehensive information on advanced synthesis, processing, characterization, manufacturing and applications in a broad range of interdisciplinary fields in science, engineering and technology.

This series accepts both authored and edited works, including textbooks, monographs, reference works, and professional books. The books in this series will provide a deep insight into the state-of-art of Materials Horizons and serve students, academic, government and industrial scientists involved in all aspects of materials research.

Review Process

The proposal for each volume is reviewed by the following:

1. Responsible (in-house) editor
2. One external subject expert
3. One of the editorial board members.

The chapters in each volume are individually reviewed single blind by expert reviewers and the volume editor.

Shikha Awasthi

Smart Hydrogels and Ferrogels

From Basic Fundamentals to Advanced Theranostics

Springer

Shikha Awasthi
Jagran School of Pharmacy
Jagran Lakecity University
Bhopal, Madhya Pradesh, India

ISSN 2524-5384 ISSN 2524-5392 (electronic)
Materials Horizons: From Nature to Nanomaterials
ISBN 978-981-95-9076-6 ISBN 978-981-95-9077-3 (eBook)
https://doi.org/10.1007/978-981-95-9077-3

This Springer imprint is published by the registered company Springer Nature Singapore Pte Ltd.
The registered company address is: 152 Beach Road, #21-01/04 Gateway East, Singapore 189721, Singapore

Contents

Chapter 1
An Introductory Chapter

1.1 Basic Concepts of Smart Hydrogels and Ferrogels

1.1.1 Hydrogels

Hydrogels are crosslinked hydrophilic polymer matrices arranged in three dimensions (3D) with a large amount of water and exhibit important properties including toughness, stretchability, deformability, softness, and biocompatibility. The crosslinked network of hydrogels prevents their immediate dissolution in the aqueous environment, and maintain their structural integrity [1, 2]. Their combination of softness and flexibility, together with their capacity to retain and maintain large amounts of water or biological fluids, makes them attractive materials. They also exhibit unique mechanical and optical qualities, biocompatibility, and remarkable physicochemical qualities like swelling and permeability, which have made them versatile tools for applications like imaging and diagnostics. Different processes, including three-dimensional (3D) polymerization, grafting polymerization, direct crosslinking of different water-soluble polymers, and radiation methods, can be used to create these interdisciplinary adaptable hydrogels. Hydrogels are created in polymerization processes with the assistance of a crosslinking agent and a monomer [3]. On the other hand, in the direct process, UV light or a chemical reaction produces free radicals [4]. Grafting polymerization involves the activation of the polymer chain using chemical means or high-energy radiation such as electron or gamma rays. This method was used to manufacture starch-based hydrogels for the transport of vitamins or drugs into the human body [5, 6]. Since no chemicals are used in the creation of the hydrogels, the polymers produced by the radiation method are the most biocompatible. The main factors influencing the applicability of synthesized hydrogels are their interfacial bonding, mechanical strength, wear resistance, and biocompatibility [7, 8]. To improve hydrogels' load-bearing capacity, strength, bioactivity, and interfacial interactions, a range of reinforcing agents are needed.

S. Awasthi, *Smart Hydrogels and Ferrogels*, Materials Horizons: From Nature to Nanomaterials, https://doi.org/10.1007/978-981-95-9077-3_1

Hydrogels may be molded into any desired form, length, size, or architecture [9], or they can be produced as thin films [10], depending on the needs. The hydrophilic functional groups like –OH, –COOH, $-NH_2$, SO_3H, etc., are responsible for the water-absorbing capabilities of hydrogels. These hydrogel networks are constructed using various polymers, either synthetic or natural, for numerous sophisticated uses. The result is a hybrid structure that may be used to sustain, shield, or connect new functions to the structure of the hydrogel. The majority of the polymers used to make hydrogels are non-toxic, appropriate for a wide range of biomedical uses, including implants and wound patches, and have been effectively employed across many industries [11–13]. Many hydrogel-based products have been used in the medical industry recently to treat patients. For instance, hydrogels based on polysaccharides have been widely used as wound dressings [14, 15], poly(2-hydroxyethyl methacrylate) [p(HEMA)] are often used in contact lens manufacturing [16], drug delivery [17] by using hydrogels containing hyaluronic acid (HA), tissue engineering or scaffolds based on proteins (gelatin, collagen, and fibrin) [18, 19]. In order to limit the toxicity of the medications to the specific location where tumor cells remain, the pharmaceuticals used to treat malignancies or tumors are directly injected into the site of the tumors or nearby areas while encapsulated in a hydrogel [20]. Stimulus-responsive hydrogels, which are regulated by varying experimental parameters like as pH, temperature, change in surface morphology, and other biological variables, have shown promise for multidimensional applications [21, 22].

1.1.2 Ferrogels

The ability of a material to perceive and react to external stimuli (magnetic fields, temperature, pH, and electric field) is the basic definition of smart materials. The polymeric structures that can respond to external magnetic fields are also known as advanced magnetic soft gels [23]. A polymeric network packed with magnetite Fe_3O_4 and other magnetic nanoparticles is the primary ingredient in ferrogels. With a high level of biocompatibility, iron (Fe) is a dietary element essential to human health. It is also a crucial element in a wide range of metabolic processes. It has been shown that adding Fe elements to gel can suit a range of demands and offer several therapeutic effects. In the hydrogel-based artificial muscle experiment, Fe ions and other building structures can increase the gel's strength, and facilitate self-healing properties [24]. Additionally, Fe^{2+} or Fe^{3+} can be reduced to Fe to induce a Fenton reaction in bacterial or tumor microenvironments. During surgical adjuvant treatment, the hydroxyl radical (·OH) generated during the reaction can efficiently destroy the cancer cells and infections [25]. In addition to its many other uses in tissue engineering, wound healing, and adjuvant therapy following the excision of malignant tumors, iron exhibits photothermal characteristics [26]. Fe-based hydrogel, for example, has very low manufacture expenses because of its simple preparation and readily available raw materials. Researchers are very interested in iron gels because of the aforementioned benefits of interdisciplinary applications. Under an electric field, the electro-active polymers (polyaniline, polypyrrole, etc.) alter in size and form. Ferrogels, in contrast to electro-active polymers, may change their physical properties in response to a magnetic field without having a physical connection to the substance.

Ferrogels' non-contact nature has therefore received a great attention from sectors where smart materials and control input are not required to come into contact [27–29]. As biosensors, ferrogels sense biological responses in various ways: ferrogels deprived of the addition of bio-receptors [30] (ionic gels sensitive to temperature, pH, and electrical fields) and ferrogels with the addition of bio-receptors (polyacrylamide) [31, 32]. The mass-based and optical biosensors use magnetoelastic, surface acoustic wave, luminescence-absorbance, fiber optic and piezoelectric techniques, while the electrochemical biosensors use conductometric, surface charge, potentiometric, impedimetric, and amperometric techniques to identify the biomolecules. Biosensors are able to detect biological reactions and translate them into signals that can be measured. Additionally, the four distinct devices that make up the biosensors are the transducer, electronic system, bio-receptor, and base material [33]. Glass, composites, metals, and polymers are essentially accepted as basic materials [34]. There are two kinds of bio-receptors: molecular species, such as genes, enzymes, antibodies, or nucleic acids and biological systems, such as cells or organs. The significant quantity of signals generated by the transducer is ascribed to the bio-receptor-analyte interaction [35]. Many attempts have been undertaken to produce theranostic materials since Funkhouser coined the term "theranostic" in 1998 to refer to a system that combines therapy and diagnostics [36]. Under the impact of external stimuli, the affected area is targeted by active or passive targeting nanostructures, releasing therapeutic medicines with negligible side effects [37, 38]. Due to the exceptional physicochemical features of Fe_3O_4, including a higher saturation field, additional anisotropy contribution, and superparamagnetism, increased field irreversibility, the gel containing superparamagnetic iron oxide nanoparticles attracted a lot of interest [39]. Under external magnetic fields, the nanosystems target the disease site, and when the external magnets are removed, the systems no longer exhibit magnetic interaction.

Ferrogels based on iron and nickel serve the biosensing and drug delivery applications [40]. While theranostic nanoplatforms for biomedical applications are thought to be an effective treatment for serious illnesses like cancer, heart disease, and AIDS [41], there are still a number of obstacles in the way of designing theranostic platforms for medical complications detection and management. For example: (i) choosing a combined theranostic nanomaterial with the right technology for the variety of therapeutic agents for the specific disease, metallic nanoparticles, and contrast imaging material becomes necessary. (ii) A highly high degree of selectivity toward the intended location is required for the suggested theranostic. (iii) Better loading and delivery productivity and imaging accuracy are prerequisites for nanotheranostics. (iv) The substances utilized in the creation of theranostics must be non-toxic, biocompatible, and readily excreted from the human body. (v) The entire nanotheranostics strategy must be inexpensive and free of adverse effects.

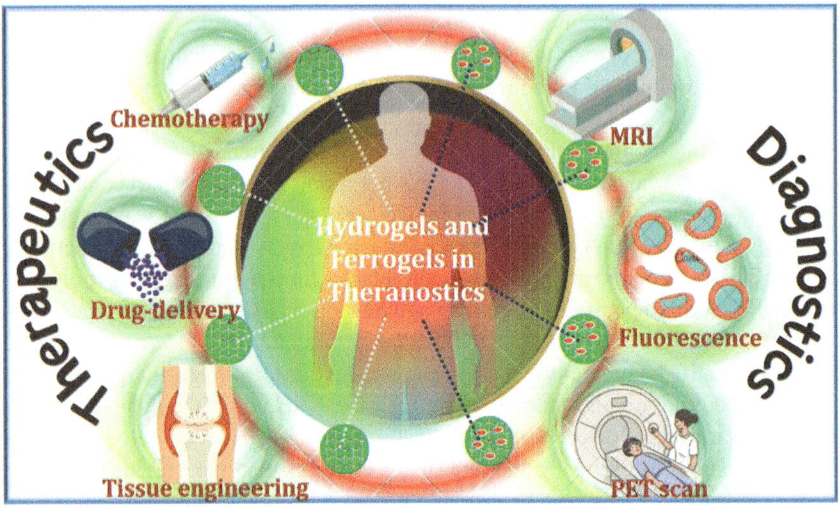

Fig. 1.1 Schematic presentation of theranostic applications of hydrogels and ferrogels

Although hydrogels and ferrogels (HG-FGs) possess a wide range of properties and limitations, these systems may all be broadly classified as hydrophilic polymeric networks. Because the HG-FGs can hold up to 99% of their weight in water, they are more suited to biological environments that are supplied with water, such as the human body [42]. The remarkable characteristics of HG-FGs provide important contributions to a range of applications (Fig. 1.1). In the world of biomedicine, hydrogels have made soft contact lenses a reality for millions of people, improving their quality of life. HG-FGs have been shown in early investigations to influence the controlled release of bioactive materials [43, 44], as well as drug delivery [45–47]. The theranostics nanoplatform for HG-FGs is schematically presented in Fig. 1.1, along with potential functionalization and therapeutic and diagnostic applications that include gene therapy, chemotherapy, phototherapy, MRI, PET scan, and fluorescence, respectively.

1.2 History of Hydrogels and Ferrogels

1.2.1 Scientific Background of Hydrogels

From the invention of colloidal gel (in 1894) to the creation of the first conventional hydrogel in 1960, researchers have been paying more attention to these materials for a variety of in vitro and in vivo theranostic applications [48]. From the 1970s forward, hydrogel research has been improving, and throughout time, new materials, reinforcements, characterizations, and applications have been developed. The history of hydrogels, according to Buwalda et al. [49], may be divided into three generations. The first generation primarily deals with the chemical adjustments made to a polymeric system of hydrogels by altering initiators, monomers and polymers, synthesis procedures, etc. The primary focus was on creating hydrogels with ideal mechanical and swelling characteristics. The second generation of hydrogels focuses on adjusting variables such as concentration, pH, and temperature to optimize the material for in situ pore creation, polymerization, or drug administration. Finally, research on stereo-complex materials is being developed by third-generation hydrogels [50].

During the nineteenth century, hydrogels were originally mentioned in relation to colloidal gels made from inorganic salts. Nevertheless, it was not until Wichterle and Lim in 1954 invented the first hydrogel (synthetic) on crosslinked poly-2-hydroxyethylmethacrylate (pHEMA) that the application of gels gained prominence [48]. Since then, the 3D structure of hydrophilic polymers has been referred to as "hydrogel" regularly. In Western Europe, soft contact lenses designed using a crosslinked macromolecular network of pHEMA's were commercially available in the 1960s. Subsequently, after successful trials, the Food and Drug Administration (FDA) authorized the pHEMA-designed lenses in 1971 [51]. The gels based on pHEMA were then used in the controlled release of drugs [52]. Updike and Hicks developed sensors in 1967 by encasing the enzyme glucose-oxidase in a hydrogel based on polyacrylamide (PAM) [53]. A schematic representation of the notable expansion of hydrogels over time is presented in Fig. 1.2. When many uses of hydrogels for the timely release of drugs, bone tissue engineering, dermatology, food items, and many other biomedical applications were described in the 1960s, the chemistry of hydrogels saw a major advance. To increase biocompatibility, novel hydrogel ideas were investigated in the 1970s, and materials including vinyl acetate, N-vinylpyrrolidone, and acrylamides were used. Tanaka did tests on PAM gels at the same period and observed that the hydrogels inclined to collapse when the solvent alignment or temperature changed. He anticipated the appearance of critical endpoints in the phase equilibria and used mean-field theory to explain the phenomena [54].

The creation of temperature-responsive hydrogels using various polymers, such as polyvinyl alcohol (PVA), polyethylene glycol (PEG), poly(N-isopropylacrylamide), etc., was the focus of increased study in the 1990s. These hydrogels were also used to create delivery systems that allow molecules to be released intermittently under the lower critical solution temperature (LCST) [55]. Injectable hydrogels were extensively studied for healthcare research in the late 1990s; Elisseeff et al. reported the first use of these hydrogels for targeted drug delivery to tissue, utilizing cartilage as a model system [56]. To enhance the mechanical strength of hydrogels, physicochemical techniques and structural modifications gained traction in the twenty-first century. These techniques included the development of homogenous networks and stereo-complex materials, the use of natural and synthetic polymers, the incorporation of inorganic materials, and the incorporation of various crosslinkers [57, 58]. Gong et al. [59] created the first novel double-network hydrogels in 2003 using a free radical polymerization process. These hydrogels showed remarkable mechanical strength,

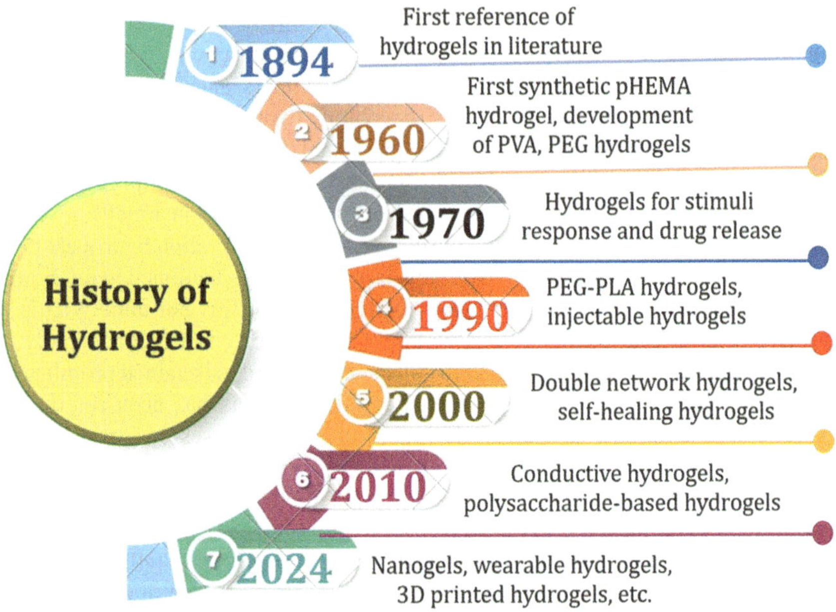

Fig. 1.2 Hydrogel advancements throughout time

like 17.2 MPa of fracture compression stress and 92% strain. Poly(2-acrylamido-2-methylpropanesulfonic acid) was used to build the first network, while PAM was used to form the second. Researchers have been able to create biocompatible hydrogels for vital biomedical applications, such as tissue engineering, contact lenses, biosensors, wound healing, etc., by achieving advancements in the synthesis of polymers (hydrophilic and hydrophobic) with a variety of structural combinations. The field of hydrogel research has rapidly evolved in recent times, with increased efforts directed toward creating novel hydrogels such as injectable hydrogels, conducting hydrogels, self-healing hydrogels, biodegradable hydrogels, wearable hydrogel patches, multicomponent hydrogels, etc. [60–62].

The proliferation of research investigations conducted in the previous 60 years (Fig. 1.3) indicates the fervor around this technology. By 2023, there will be over 100,000 papers about hydrogels, a huge rise from the mere 1000 publications in 1980. Only half of 2024s papers based on hydrogels are indexed by the "Web of Science," totaling over 1000.

1.2.2 History of Ferrogels

Magnetic gels are typically polyelectrolyte gels and are members of the hydrogel and polymeric gel families. These are crosslinked, extremely inflated molecular networks

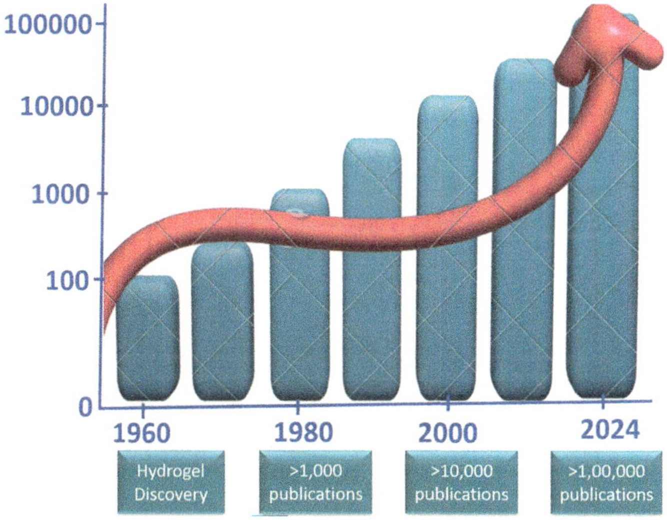

Fig. 1.3 According to statistics from the Web of Science, the number of papers referencing hydrogels has grown rapidly since its creation in 1960

that solidify into a hydrophilic state. The initial reports of reversible swelling and deswelling of polymeric gels by ionization were from Kuhn et al. [63] and Katchalsky et al. [64]. Afterward, it was demonstrated by Tanaka et al. [65] that these gels may collapse in an electric field. As confirmed later by Osada and the research team [66], De Rossi et al. [67], and Schneider and co-workers [68], it is also possible to cause this type of swelling and deswelling chemically (pH Muscles) or electrically (EAP). Zrinyi et al. developed the first magnetically active gels with responses that can be significantly accelerated by an applied magnetic field [69–71]. Joseph and Ronald created the first ferrohydrodynamic materials, or ferrogels, in 1964 [72]. Following that, ferrogels emerged as a promising option for a variety of uses, mostly in therapeutic settings. Cell tracking, hypothermal therapy, magnetic resonance imaging (MRI), magnetic biosensing, and nanotheranostic advancements are just a few of the numerous biomedical uses [73–76].

The abundance of studies carried out during the last half-century (Fig. 1.4) demonstrates the enthusiasm around this technology. The number of papers discussing ferrogels will reach thousands by 2023, a significant increase from the meager 50 publications in 2000. The "Web of Science," which has over 250 entries, only has half of the publications published in 2024 that use ferrogels indexed.

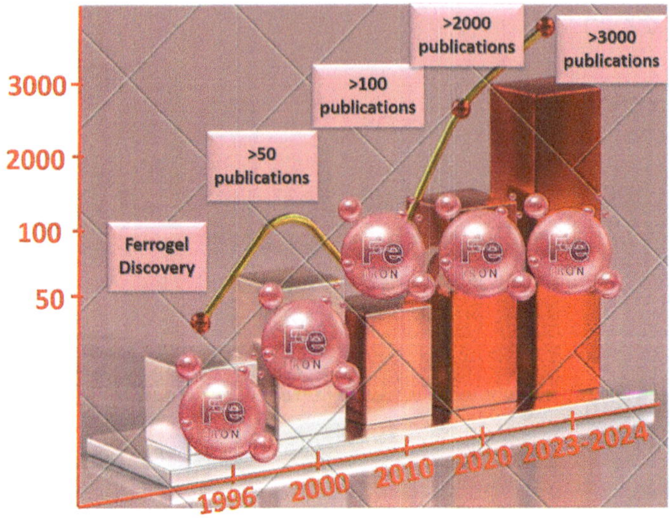

Fig. 1.4 Since the invention of ferrogels in 1996, the number of publications mentioning them has increased significantly, based on data from the Web of Science

1.3 Summary and Research Directions

The chapter clarifies the characteristics of HG-FGs in relation to their composition, manufacturing processes, ongoing studies, and uses in various applications. A description of HG-FGs reveals that they have a polymeric structure with potential uses as a substrate for nanotheranostics. The polymeric network can be created by free radical polymerization by combining a suitable quantity of monomer with the crosslinker. Using metal salt and a reducing agent, the ions were reduced to embed the nanoparticles in the HG-FGs. Consequently, the scientific community may benefit greatly from such benchmark studies on the synthesis of hybrid HG-FGs based on nanoparticles, since they provide an in-depth synopsis of the subject. The author firmly believes that this report will garner widespread attention from the creative community in the fields of chemicals, biology, and materials because it is meant to be a comprehensive and critical analysis of general interest for engineers or scientists working on smart hydrogels and ferrogels, as well as for the benefit of a worldwide readership (beginners as well as non-experts).

References

1. J.Y. Sun, X. Zhao, W.R.K. Illeperuma, O. Chaudhuri, K.H. Oh, D.J. Mooney, J.J. Vlassak, Z. Suo, Highly stretchable and tough hydrogels. Nature **489**, 133–136 (2012)
2. R. Zhong, S. Talebian, B.B. Mendes, G. Wallace, R. Langer, J. Conde, J. Shi, Hydrogels for RNA delivery. Nat. Mater. **22**, 818–831 (2023). https://doi.org/10.1038/s41563-023-01472-w

3. S. Awasthi, S.K. Pandey, Recent advances in smart hydrogels and carbonaceous nanoallotropes composites. Appl. Mater. Today **36**, 102058 (2024). https://doi.org/10.1016/j.apmt.2024.102058
4. T. Funami, M. Hiroe, S. Noda, I. Asai, S. Ikeda, K. Nishinari, Influence of molecular structure imaged with atomic force microscopy on the rheological behavior of carrageenan aqueous systems in the presence or absence of cations. Food Hydrocoll. **21**, 617–629 (2007)
5. L.S. Spinelli, A.S. Aquino, E. Lucas, A.R. D'Almeida, R. Leal, A.L. Martins, M. In, Adsorption of polymers used in drilling fluids on the inner surfaces of carbon steel pipes. Polym. Eng. Sci. **48**, 1885–1891 (2008)
6. M. Zhai, F. Yoshii, T. Kume, K. Hashim, Syntheses of PVA/starch grafted hydrogels by irradiation. Carbohydr. Polym. **50**, 295–303 (2002)
7. A. Kumar, A. Sood, R. Singhmar, Y.K. Mishra, V.K. Thakur, S.S. Han, Manufacturing functional hydrogels for inducing angiogenic–osteogenic coupled progressions in hard tissue repairs: prospects and challenges. Biomater. Sci. **10**, 5472–5497 (2022)
8. R. Mohammadinejad, H. Maleki, E. Larrañeta, A.R. Fajardo, A.B. Nik, A. Shavandi, A. Sheikhi, M. Ghorbanpour, M. Farokhi, P. Govindh, E. Cabane, S. Azizi, A.R. Aref, M. Mozafari, M. Mehrali, S. Thomas, J.F. Mano, Y.K. Mishra, V.K. Thakur, Status and future scope of plant-based green hydrogels in biomedical engineering. Appl. Mater. Today **16**, 213–246 (2019)
9. Z. You, M. Behl, S.L. Grage, J. Bürck, Q. Zhao, A.S. Ulrich, A. Lendlein, Shape-memory effect by sequential coupling of functions over different length scales in an architectured hydrogel. Biomacromolecules **21**, 680–687 (2020). https://doi.org/10.1021/acs.biomac.9b01390
10. X. Zhang, Y. Guan, Y. Zhang, Ultrathin hydrogel films for rapid optical biosensing. Biomacromolecules **13**, 92–97 (2012). https://doi.org/10.1021/bm2012696
11. C. Chen, Y. Wang, H. Zhang, H. Zhang, W. Dong, W. Sun, Y. Zhao, Responsive and self-healing structural color supramolecular hydrogel patch for diabetic wound treatment. Bioact. Mater. **15**, 194–202 (2022). https://doi.org/10.1016/j.bioactmat.2021.11.037
12. A. Choe, J. Yeom, R. Shanker, M.P. Kim, S. Kang, H. Ko, Stretchable and wearable colorimetric patches based on thermoresponsive plasmonic microgels embedded in a hydrogel film. NPG Asia Mater. **10**, 912–922 (2018). https://doi.org/10.1038/s41427-018-0086-6
13. Z. Jing, R. Ni, J. Wang, X. Lin, D. Fan, Q. Wei, T. Zhang, Y. Zheng, H. Cai, Z. Liu, Practical strategy to construct anti-osteosarcoma bone substitutes by loading cisplatin into 3D-printed titanium alloy implants using a thermosensitive hydrogel. Bioact. Mater. **6**, 4542–4557 (2021). https://doi.org/10.1016/j.bioactmat.2021.05.007
14. M. Zhang, M. Yang, M.W. Woo, Y. Li, W. Han, X. Dang, High-mechanical strength carboxymethyl chitosan-based hydrogel film for antibacterial wound dressing. Carbohydr. Polym. **256**, 117590 (2021). https://doi.org/10.1016/j.carbpol.2020.117590
15. D.Z. Zmejkoski, N.M. Zdravković, D.D. Trišić, M.D. Budimir, Z.M. Marković, N.O. Kozyrovska, B.M. Todorović Marković, Chronic wound dressings—pathogenic bacteria anti-biofilm treatment with bacterial cellulose-chitosan polymer or bacterial cellulose-chitosan dots composite hydrogels. Int. J. Biol. Macromol. **191**, 315–323 (2021). https://doi.org/10.1016/j.ijbiomac.2021.09.118
16. P.C. Nicolson, J. Vogt, Soft contact lens polymers: an evolution. Biomaterials **22**, 3273–3283 (2001). https://doi.org/10.1016/S0142-9612(01)00165-X
17. S. Ekici, P. Ilgin, S. Butun, N. Sahiner, Hyaluronic acid hydrogel particles with tunable charges as potential drug delivery devices. Carbohydr. Polym. **84**, 1306–1313 (2011). https://doi.org/10.1016/j.carbpol.2011.01.028
18. S. Sharifi, M.M. Islam, H. Sharifi, R. Islam, D. Koza, F. Reyes-Ortega, D. Alba-Molina, P.H. Nilsson, C.H. Dohlman, T.E. Mollnes, J. Chodosh, M. Gonzalez-Andrades, Tuning gelatin-based hydrogel towards bioadhesive ocular tissue engineering applications. Bioact. Mater. **6**, 3947–3961 (2021). https://doi.org/10.1016/j.bioactmat.2021.03.042
19. G. Montalbano, S. Toumpaniari, A. Popov, P. Duan, J. Chen, K. Dalgarno, W.E. Scott, A.M. Ferreira, Synthesis of bioinspired collagen/alginate/fibrin based hydrogels for soft tissue engineering. Mater. Sci. Eng. C **91**, 236–246 (2018). https://doi.org/10.1016/j.msec.2018.04.101

20. Y.-W. Lo, M.-T. Sheu, W.-H. Chiang, Y.-L. Chiu, C.-M. Tu, W.-Y. Wang, M.-H. Wu, Y.-C. Wang, M. Lu, H.-O. Ho, In situ chemically crosslinked injectable hydrogels for the subcutaneous delivery of trastuzumab to treat breast cancer. Acta Biomater. **86**, 280–290 (2019). https://doi.org/10.1016/j.actbio.2019.01.003
21. S. Dutta, P. Samanta, D. Dhara, Temperature, pH and redox responsive cellulose based hydrogels for protein delivery. Int. J. Biol. Macromol. **87**, 92–100 (2016). https://doi.org/10.1016/j.ijbiomac.2016.02.042
22. P. Abdollahiyan, B. Baradaran, M. de la Guardia, F. Oroojalian, A. Mokhtarzadeh, Cutting-edge progress and challenges in stimuli responsive hydrogel microenvironment for success in tissue engineering today. J. Control. Release **328**, 514–531 (2020). https://doi.org/10.1016/j.jconrel.2020.09.030
23. Magnetic gel delivers drugs. Nature **468**, 1004–1004 (2010). https://doi.org/10.1038/4681004c
24. K. Roemhild, F. von Maltzahn, R. Weiskirchen, R. Knüchel, S. von Stillfried, T. Lammers, Iron metabolism: pathophysiology and pharmacology. Trends Pharmacol. Sci. **42**, 640–656 (2021). https://doi.org/10.1016/j.tips.2021.05.001
25. Y. Hao, Z. Dong, M. Chen, Y. Chao, Z. Liu, L. Feng, Y. Hao, Z.L. Dong, M.C. Chen, Y. Chao, Z. Liu, L.Z. Feng, Near-infrared light and glucose dual-responsive cascading hydroxyl radical generation for in situ gelation and effective breast cancer treatment. Biomaterials **228**, 119568 (2020)
26. J. Wang, L. Wang, Y. Gao, Z. Zhang, X. Huang, T. Han, B. Liu, Y. Zhang, Y. Li, L. Zhang, Synergistic therapy of celecoxib-loaded magnetism-responsive hydrogel for tendon tissue injuries. Front. Bioeng. Biotechnol. **8**, 592068–592076 (2020). https://doi.org/10.3389/fbioe.2020.592068
27. S. Sturm, M. Siglreitmeier, D. Wolf, K. Vogel, M. Gratz, D. Faivre, A. Lubk, B. Büchner, E.V. Sturm, H. Cölfen, Magnetic nanoparticle chains in gelatin ferrogels: bioinspiration from magnetotactic bacteria. Adv. Funct. Mater. **29**, 1905996 (2019). https://doi.org/10.1002/adfm.201905996
28. M. Helminger, B. Wu, T. Kollmann, D. Benke, D. Schwahn, V. Pipich, D. Faivre, D. Zahn, H. Cölfen, Synthesis and characterization of Gelatin-based magnetic hydrogels. Adv. Funct. Mater. **24**, 3187–3196 (2014). https://doi.org/10.1002/adfm.201303547
29. S.-H. Hu, T.-Y. Liu, D.-M. Liu, S.-Y. Chen, Controlled pulsatile drug release from a ferrogel by a high-frequency magnetic field. Macromolecules **40**, 6786–6788 (2007). https://doi.org/10.1021/ma0707584
30. H.J. Hwang, M.Y. Ryu, C.Y. Park, J. Ahn, H.G. Park, C. Choi, S. Do Ha, T.J. Park, J.P. Park, High sensitive and selective electrochemical biosensor: label-free detection of human norovirus using affinity peptide as molecular binder. Biosens. Bioelectron. **87**, 164–170 (2017)
31. F. Blyakhman, N. Buznikov, T. Sklyar, A. Safronov, E. Golubeva, A. Svalov, S. Sokolov, G. Melnikov, I. Orue, G. Kurlyandskaya, Mechanical, electrical and magnetic properties of ferrogels with embedded iron oxide nanoparticles obtained by laser target evaporation: focus on multifunctional biosensor applications. Sensors **18**, 872 (2018). https://doi.org/10.3390/s18030872
32. A. Safronov, E. Mikhnevich, Z. Lotfollahi, F. Blyakhman, T. Sklyar, A. Larrañaga Varga, A. Medvedev, S. Fernández Armas, G. Kurlyandskaya, Polyacrylamide ferrogels with magnetite or strontium hexaferrite: next step in the development of soft biomimetic matter for biosensor applications. Sensors **18**, 257 (2018). https://doi.org/10.3390/s18010257
33. J. Tavakoli, Y. Tang, Hydrogel based sensors for biomedical applications: an updated review. Polymers (Basel) **9**, 1–25 (2017)
34. G. Ertürk, B. Mattiasson, Molecular imprinting techniques used for the preparation of biosensors. Sensors (Switzerland) **17**, 1–17 (2017)
35. T. Vo-Dinh, Nanobiosensors: probing the sanctuary of individual living cells. J. Cell. Biochem. 154–161 (2002)
36. J. Funkhouser, Reinventing pharma: the theranostic revolution. Curr. Drug Discov. **2**, 17–19 (2002)

37. V.J. Yao, S. D'Angelo, K.S. Butler, C. Theron, T.L. Smith, S. Marchiò, J.G. Gelovani, R.L. Sidman, A.S. Dobroff, C.J. Brinker, A.R.M. Bradbury, W. Arap, R. Pasqualini, Ligand-targeted theranostic nanomedicines against cancer. J. Control. Release **240**, 267–286 (2016). https://doi.org/10.1016/j.jconrel.2016.01.002
38. M. Swierczewska, H.S. Han, K. Kim, J.H. Park, S. Lee, Polysaccharide-based nanoparticles for theranostic nanomedicine. Adv. Drug Deliv. Rev. **99**, 70–84 (2016). https://doi.org/10.1016/j.addr.2015.11.015
39. A. Farnoudian-Habibi, S. Kangari, B. Massoumi, M. Jaymand, Determination of losartan potassium in the presence of hydrochlorothiazide via a combination of magnetic solid phase extraction and fluorometry techniques in urine samples. RSC Adv. **5**, 102895–102903 (2015). https://doi.org/10.1039/C5RA20117A
40. J. Li, K.P. Lee, A.I. Gopalan, One-step preparation of nickel nanoparticle-based magnetic poly(vinyl alcohol) gels. Coatings **9**, 744–756 (2019)
41. H. Cui, J. Wang, Progress in the development of nanotheranostic systems. Theranostics **6**, 915–917 (2016). https://doi.org/10.7150/thno.16153
42. Q. Wang, Y. Zhang, Y. Ma, M. Wang, G. Pan, Nano-crosslinked dynamic hydrogels for biomedical applications. Mater. Today Bio **20**, 100640 (2023)
43. P. Andrade-Vivero, E. Fernandez-Gabriel, C. Alvarez-Lorenzo, A. Concheiro, Improving the loading and release of NSAIDs from pHEMA hydrogels by copolymerization with functionalized monomers. J. Pharm. Sci. **96**, 802–813 (2007)
44. N.B. Graham, M.E. Mcneiu, Hydrogels for controlled drug delivery. Biomaterials **5**, 27–36 (1984)
45. R.W. Korsmeyer, N.A. Peppas, Effect of the morphology of hydrophilic polymeric matrices on the diffusion and release of water soluble drugs. J. Membr. Sci. **9**, 211–227 (1981)
46. R.A. Haldon, B.E. Leeb, Structure and permeability of porous films of poly(hydroxy ethyl methacrylate). Br. Polym. J. **4**, 491–501 (1972)
47. B. Gupta, R. Agarwal, M. Alam, *Hydrogels for Wound Healing Applications. Biomedical Hydrogels: Biochemistry, Manufacture and Medical Applications* (Woodhead Publishing, New Delhi, 2011)
48. O. Wichterle, D. Lím, Hydrophilic gels for biological use. Nature **185**, 117–118 (1960). https://doi.org/10.1038/185117a0
49. S.J. Buwalda, K.W.M. Boere, P.J. Dijkstra, J. Feijen, T. Vermonden, W.E. Hennink, Hydrogels in a historical perspective: from simple networks to smart materials. J. Control. Release **190**, 254–273 (2014). https://doi.org/10.1016/j.jconrel.2014.03.052
50. H.J. Chung, Y. Lee, T.G. Park, Thermo-sensitive and biodegradable hydrogels based on stereocomplexed pluronic multi-block copolymers for controlled protein delivery. J. Control. Release **127**, 22–30 (2008)
51. C. Maldonado-Codina, N. Efron, Impact of manufacturing technology and material composition on the mechanical properties of hydrogel contact lenses. Ophthalmic Physiol. Opt. **24**, 551–561 (2004). https://doi.org/10.1111/j.1475-1313.2004.00236.x
52. W.E. Roorda, M.A. de Vries, L.G.J. de Leede, A.G. de Boer, D.D. Breimer, H.E. Junginger, Zero-order release of oxprenolol-HCl, a new approach. J. Control. Release **7**, 45–52 (1988). https://doi.org/10.1016/0168-3659(88)90079-X
53. S.J. Updike, G.P. Hicks, The enzyme electrode. Nature **214**, 986–988 (1967). https://doi.org/10.1038/214986a0
54. T. Tanaka, Collapse of gels and the critical endpoint. Phys. Rev. Lett. **40**, 820–823 (1978). https://doi.org/10.1103/PhysRevLett.40.820
55. R. Dinarvand, A. D'Emanuele, The use of thermoresponsive hydrogels for on-off release of molecules. J. Control. Release **36**, 221–227 (1995). https://doi.org/10.1016/0168-3659(95)00035-7
56. J. Elisseeff, K. Anseth, D. Sims, W. McIntosh, M. Randolph, R. Langer, Transdermal photopolymerization for minimally invasive implantation. Proc. Natl. Acad. Sci. **96**, 3104–3107 (1999). https://doi.org/10.1073/pnas.96.6.3104

57. X. Jia, J.A. Burdick, J. Kobler, R.J. Clifton, J.J. Rosowski, S.M. Zeitels, R. Langer, Synthesis and characterization of in situ cross-linkable hyaluronic acid-based hydrogels with potential application for vocal fold regeneration. Macromolecules **37**, 3239–3248 (2004). https://doi.org/10.1021/ma035970w
58. X. Jia, K.L. Kiick, Hybrid multicomponent hydrogels for tissue engineering. Macromol. Biosci. **9**, 140–156 (2009). https://doi.org/10.1002/mabi.200800284
59. J.P. Gong, Y. Katsuyama, T. Kurokawa, Y. Osada, Double-network hydrogels with extremely high mechanical strength. Adv. Mater. **15**, 1155–1158 (2003). https://doi.org/10.1002/adma.200304907
60. F. Huang, W. Wei, Q. Fan, L. Li, M. Zhao, Z. Zhou, Super-stretchable and adhesive cellulose nanofiber-reinforced conductive nanocomposite hydrogel for wearable motion-monitoring sensor. J. Colloid Interface Sci. **615**, 215–226 (2022). https://doi.org/10.1016/j.jcis.2022.01.117
61. J. Shi, L. Yu, J. Ding, PEG-based thermosensitive and biodegradable hydrogels. Acta Biomater. **128**, 42–59 (2021). https://doi.org/10.1016/j.actbio.2021.04.009
62. C. Lim, Y.J. Hong, J. Jung, Y. Shin, S.-H. Sunwoo, S. Baik, O.K. Park, S.H. Choi, T. Hyeon, J.H. Kim, S. Lee, D.-H. Kim, Tissue-like skin-device interface for wearable bioelectronics by using ultrasoft, mass-permeable, and low-impedance hydrogels. Sci. Adv. **7** (2021). https://doi.org/10.1126/sciadv.abd3716
63. W. Kuhn, B. Hargitay, A. Katchalsky, H. Eisenberg, Reversible dilation and contraction by changing the state of ionization of high-polymer acid networks. Nature **165**, 514–516 (1950). https://doi.org/10.1038/165514a0
64. A. Katchalsky, Rapid swelling and deswelling of reversible gels of polymeric acids by ionization. Experientia **5**, 319–320 (1949). https://doi.org/10.1007/BF02172636
65. T. Tanaka, I. Nishio, S.-T. Sun, S. Ueno-Nishio, Collapse of gels in an electric field. Science **218**(1982), 467–469 (1979). https://doi.org/10.1126/science.218.4571.467
66. Y. Osada, Conversion of chemical into mechanical energy by synthetic polymers (chemomechanical systems), in *Polymer Physics* (Springer-Verlag, Berlin/Heidelberg, n.d.), pp. 1–46. https://doi.org/10.1007/BFb0024041
67. D. DeRossi, K. Kajiwara, Y. Osada, A. Yamauchi (eds.), *Polymer Gels* (Springer US, Boston, MA, 1991). https://doi.org/10.1007/978-1-4684-5892-3
68. H.-J. Schneider, K. Kato, R.M. Strongin, Chemomechanical polymers as sensors and actuators for biological and medicinal applications. Sensors **7**, 1578–1611 (2007). https://doi.org/10.3390/s7081578
69. M. Zrínyi, L. Barsi, A. Büki, Ferrogel: a new magneto-controlled elastic medium. Polym. Gels Netw. **5**, 415–427 (1997). https://doi.org/10.1016/S0966-7822(97)00010-5
70. M. Zrínyi, L. Barsi, A. Büki, Deformation of ferrogels induced by nonuniform magnetic fields. J. Chem. Phys. **104**, 8750–8756 (1996). https://doi.org/10.1063/1.471564
71. M. Zrinyi, F. Horkay, On the thermodynamics of chemomechanical energy conversion realised by neutral gels. J. Intell. Mater. Syst. Struct. **4**, 190–201 (1993). https://doi.org/10.1177/1045389X9300400208
72. J.L. Neuringer, R.E. Rosensweig, Ferrohydrodynamics. Phys. Fluids **7**, 1927–1937 (1964). https://doi.org/10.1063/1.1711103
73. Z. Zhang, N. Mascheri, R. Dharmakumar, Z. Fan, T. Paunesku, G. Woloschak, D. Li, Superparamagnetic iron oxide nanoparticle-labeled cells as an effective vehicle for tracking the GFP gene marker using magnetic resonance imaging. Cytotherapy **11**, 43–51 (2009). https://doi.org/10.1080/14653240802420243
74. J.S. Weinstein, C.G. Varallyay, E. Dosa, S. Gahramanov, B. Hamilton, W.D. Rooney, L.L. Muldoon, E.A. Neuwelt, Superparamagnetic iron oxide nanoparticles: diagnostic magnetic resonance imaging and potential therapeutic applications in neurooncology and central nervous system inflammatory pathologies, a review. J. Cereb. Blood Flow Metab. **30**, 15–35 (2010). https://doi.org/10.1038/jcbfm.2009.192
75. S.-H. Liao, H.-S. Huang, J.-J. Chieh, Y.-K. Su, Y.-F. Tong, K.-W. Huang, Characterizations of anti-alpha-fetoprotein-conjugated magnetic nanoparticles associated with alpha-fetoprotein for biomedical applications. Sensors **17**, 2018 (2017). https://doi.org/10.3390/s17092018

76. G.V. Kurlyandskaya, D.S. Portnov, I.V. Beketov, A. Larrañaga, A.P. Safronov, I. Orue, A.I. Medvedev, A.A. Chlenova, M.B. Sanchez-Ilarduya, A. Martinez-Amesti, A.V. Svalov, Nanostructured materials for magnetic biosensing. Biochim. Biophys. Acta Gen. Subj. **1861**, 1494–1506 (2017). https://doi.org/10.1016/j.bbagen.2016.12.003

Chapter 2
Crosslinking Approaches, Design Strategies, and Physical Properties of Hydrogels and Ferrogels

2.1 Design Mechanism of Hydrogels

Depending on the desired architecture and function, hydrogels can be synthesized through the formation of crosslinks via various techniques. The prime crosslinking methods for hydrogel synthesis involve physical and chemical crosslinking (Fig. 2.1). Other common fabrication methods include graft polymerization and radiation-induced crosslinking.

2.1.1 Chemical Crosslinking

To achieve irreversible network formation in hydrogels, the covalent crosslinking approach is employed, utilizing stable covalent bonds between polymer chains. Using crosslinking reagents or monomer conjugations with the polymer backbone is known as chemical crosslinking. Synthetic and natural polymers with functional groups like amino, carboxyl, and alcoholic groups can undergo crosslinking through interaction with crosslinking agents like aldehyde groups (e.g., adipic acid dihydrazide, glutaraldehyde, etc.). Cross-linking was produced using polymer-polymer conjugation, enzyme-catalyzed methods, and photosensitive agents [2]. Here are several techniques for carrying out chemical cross-linking.

2.1.1.1 Schiff Base Reactions: Crosslinking Between Amino and Aldehyde Groups

Crosslinking serves as a method for enhancing the stability of polymers in macromolecular chemistry that creates a three dimensional polymeric network [3]. An ionic or covalent bond between two polymer chains is called a crosslinker. Crosslinkers

S. Awasthi, *Smart Hydrogels and Ferrogels*, Materials Horizons: From Nature to Nanomaterials, https://doi.org/10.1007/978-981-95-9077-3_2

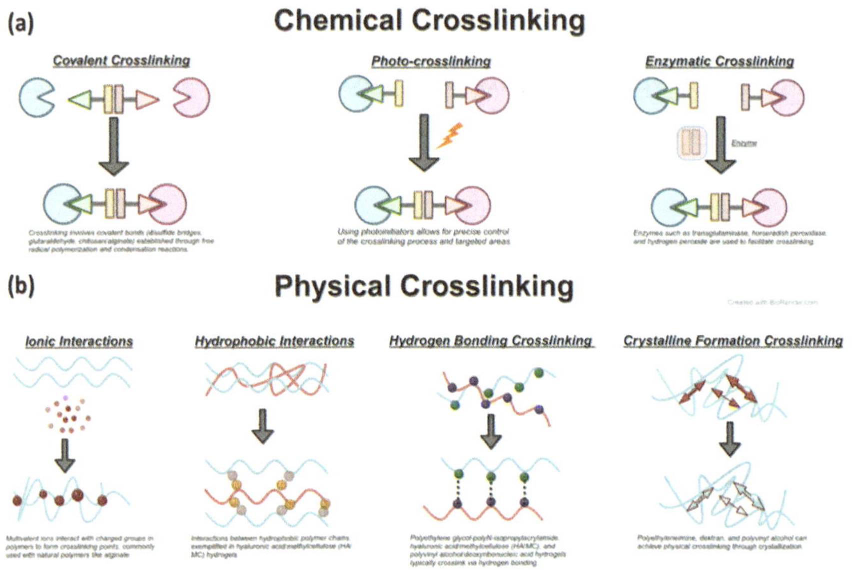

Fig. 2.1 Hydrogel crosslinking methods schematic diagram. **A** Classification of hydrogel and principles of chemical crosslinking. **B** Classification of hydrogel and principles of physical crosslinking. The tiny spheres stand in for various ions or atoms that are involved in crosslinking. Crosslinking reactions are shown by the gray arrows. Electrostatic forces are represented by red and green arrows. Reproduced with permission from Ref. [1]. Copyright 2024 MDPI (OA under the license CC-BY-NC-ND)

connect two polymeric chains by creating a bridge between their two reactive functional groups. This method involves inserting a new molecule between the polymeric strands to create a crosslinked network [4]. The most fundamental type of crosslinking occurs between the amino functional group and formyl groups when creating a Schiff base. For instance, a Schiff-base reaction is formed between the amino functional groups of chitosan and dialdehydes like glutaraldehyde and glyoxal via the covalent imine bond. Genipin, a naturally occurring compound derived from the fruit of *Gardenia*, is frequently utilized as a crosslinking agent due to its biocompatibility, which serves as an alternative to traditional dialdehyde-based crosslinkers. Moreover, it has been established that genipin undergoes covalent interactions with polymers like chitosan and gelatin [5]. A hydrogel synthesized using glutaraldehyde as a crosslinker to crosslink cornstarch and polyvinyl alcohol. The resulting hydrogel layer may serve as a synthetic skin substitute and can also transport nutrition, anti-inflammatory agents, and pharmaceuticals to the targeted sites [6].

2.1.1.2 Hybrid Polymer Networks (HPN) or Polymer-Polymer Crosslinking

In hydrogels, the structural units of two polymeric chains undergo a process known as crosslinking. Reactive functional groups, therefore, need to be introduced in the matrix polymer beforehand. Different covalent connections between the polymer chains are formed depending on the kind of functional groups, the crosslinker rate, and the degradability profile of the final product. This technique enables quick synthesis with a variety of hydrogels, and the creation of various connections within the hydrogel network [7].

2.1.1.3 Photo Crosslinking and Free Radical Polymerization

Hydrogels that are photo-crosslinked can only develop when the photosensitive functional group is present. UV light exposure can cause a polymer featuring a photosensitive functional group to develop crosslinks. Of the polymers used in photo-crosslinking, chitosan has attracted the most interest [8]. A photo-initiated chitosan hydrogel was crosslinked by introducing azide groups ($-N_3$) to the polymeric chain of chitosan. The azido group is converted to an imene group (R–N:) by exposure to UV radiation, and this group then ties the amine groups of the chitin molecule, causing the hydrogel to form quickly. It is also possible for a photo-crosslinked hydrogel to develop between the polymers. An example of this is a thermosensitive chitosan-pluronic hydrogel, in which photosensitive acrylate groups ($H_2C{=}CHCOO$) were functionalized into both polymers using UV light. Hydrogels composed of GO and PEG were produced via thiol-ene photo-polymerization [9].

2.1.1.4 Crosslinking by Chemical Reactions of Complementary Groups

Crosslinking of polymer chains may entail chemical interactions between complementary functional groups on crosslinkers and the polymers. Typical examples of "click" biochemistry include thiol-Michael addition, Schiff-base reaction, condensation reaction, azide-alkyne cycloaddition, Diels–Alder cycloaddition, and many more [10]. Certain functional groups must be incorporated into the polymeric chains for every reaction. Hydrogels have been made via the "click" reaction known as azide-alkyne cycloaddition, which occurs between an azido group and an alkyl group, either an internal or terminal [11]. Laezza et al. developed a chemical method for the semi-synthesis of an amide linker (ELAHA) hyaluronan-elastin conjugate in the presence of tris (2-carboxyethyl) phosphine hydrochloride (TCEP · HCl) to design a bioactive component in electrospun scaffolds that will be used as a wound dressing material [12]. The synthetic process known as thiol-Michael addition is widely employed in the conjugation of peptides to polysaccharides for use in biomaterials. Mitsuhashi et al. created moldable tissue-sealant hydrogels by combining the yield stress fluidity of a carbomer with in situ crosslinking of 3-arm PEG-thiol (PEG-SH) and 4-arm

PEG-acrylate (PEG-AC). Two aqueous precursors were created by combining the carbomer with each PEG oligomer: carboxypolymethylene/PEG-SH and carboxypolymethylene/PEG-AC. When placed on the tissue surface, the duplet hydrogel precursors showed enough yield strength (> 100 Pa) to prevent leakage. Furthermore, these hydrogel precursors showed quick restructuring with repeated changes in the shear strain. These hydrogel precursors' moldability is influenced by their rheological characteristics. Following the mixing of these two precursors, they underwent thiol-Michael addition to the change from yield stress fluids to hydrogels that were chemically crosslinked, or carboxypolymethylene/PEG hydrogel. In several cases, the hydrogel's biocompatibility may be adjusted via the quantity of crosslinker (DTT) employed, and that kind of hydrogel may work well for 3D cell culture and cell transport implementations [13]. The Diels–Alder cycloaddition reaction is an additional example of "click" chemistry that takes place between a dienophile group and an olefin. Two typical substituted alkene complexes are furan-maleimide and tetrazine-norbornene. The most prevalent way for producing hydrogels utilizing Diels–Alder "click" chemistry is to initially construct precursors of hydrogels by customizing polymers with vital functional groups. Later, combining the precursor mixture for hydrogel, the polymers undergo crosslinking, which produces hydrogel. For instance, Hu et al. [14] used the Diels–Alder click reaction to create an injectable hydrogel containing human umbilical cord-derived mesenchymal stem cells (UCMSCs) for endometrial regeneration and fertility restoration. The hydrogel had acceptable mechanical properties, good biocompatibility, and desired degradation properties [15].

Chemically reversible hydrogels can be formed via free radical polymerization of a polymerizable group in a hydrophilic polymer. Radical polymerization is a widely utilized conjugation method for creating hydrogels because of its supremacy in terms of elevated reactivity, high conversion, and moderate reaction conditions [16]. Using this technique, gels were made from hydrophilic polymers that were natural, synthetic, or semi-synthetic. Free radical polymerization proceeds in three steps: initiation, propagation, and termination. Starters, also known as initiators, are added to a mixture of substituted vinyl monomers and crosslinkers to initiate polymerization by generating free radicals. Common initiators involve azobis (cyanovaleric acid) (ACVA) and ammonium persulfate (APS). As a result of the polymer chains spreading and becoming crosslinked, the solution becomes more viscous, ultimately causing gelation by halting the polymerization process. Nitrogen gas is often used to cleanse the system since atmospheric oxygen functions as a scavenger of free radicals [17].

2.1.2 Physical Crosslinkage

Ionic interactions, self-assembling crosslinking, and freeze-thaw cycles are all included in physical crosslinking. Non-polar interactions among molecules, such as hydrophobic interactions, hydrogen bonding, and electrostatic forces, are necessary for self-assembling crosslinking. Hydrophobic groups, electrostatic attraction, and hydrogen bonding between molecules combine to form self assembled hydrogels, such as physically reversible hydrogels, which are then produced by non-self-assembling macromolecules. Physical crosslinking of hydrogels prevents their disintegration through physical interactions between different polymer chains [18]. Due to the physically reversible hydrogels that can return to their copolymer, these connections are disrupted by the utilization of stress or changes in the physical surroundings. Emerging materials are physically conjugated hydrogels because they are easy to synthesize and do not require a crosslinker. The stability of the encapsulated materials (such as protein cells, etc.) is impacted by these crosslinking agents [18]. Below is a discussion of some physical interactions related to the amalgamation of physically crosslinked hydrogels.

2.1.2.1 Interaction of Ions

Ionic polymers are crosslinked using this method. Ionic polymers, for example, can be crosslinked by the introduction of divalent or trivalent counter-ions [19]. The principle behind this method is to gel a polymeric electrolyte solution (like sodium alginate) with multivalent particles that have the opposite charge (like $Ca^{2+} + 2Cl^{-}$).

2.1.2.2 Complex Coacervation

A polyatomic anion and a polyatomic cation polymer can combine to form complex coacervate gels. This process creates soluble and insoluble complexes by the adhesion of the opposing polymer charges. These complexes depend on the pH of the solution and the amount present [20]. One example is the coacervation of soybean glycinin (11S)-chitosan (CS) complex gels with a range of textural characteristics. The procedure included heating the 11S-CS coacervates to produce the final complex gels after they had first formed by electrostatic interactions [21].

2.1.2.3 Dipole-Dipole Attraction

The created hydrogel can be regulated with temperature, solvent type, solution concentrations, and polymer functionalities. By raising the concentration of polymeric, hydrogels become more stable and organized due to increased entanglements and hydrogen (H-)bonding interactions [22]. Diluting the solution, however, weakens and disturbs the network H-bonding in a matter of hours (h). For example, the two frontiers (PAM−CNT and PAM−TiO_2) that make up the PAM−TiO_2−CNT composite model were stabilized via van der Waals-type interactions and H-bonding. The PAM−TiO_2−CNT composite has been determined to be the most stable use of NCI plot, HOMO−LUMO gap, and natural population methods. Consequently, the synthesized polyacrylamide hydrogels combined with the TiO_2 and CNT may prove to be an exceptional hydrogel nanocomposite for cartilage restoration applications [23].

Repeating the freeze and thaw procedure, that is, repeatedly freezing and then melting at room temperature, will yield hydrogels. The creation of structures containing microcrystals is a part of this process [24]. Nevertheless, this technique also depends on the creation of H-bonds between the polymer chains; PVA is the better typical and extensively researched polymer that can be transformed into a hydrogel through freezing-thawing [25].

2.1.2.4 Aggregation Induced by Heat Maturation

Here, the hydrogel is generated by delivering a high temperature that stimulates the accumulation of the polymers. Odelli et al. studied the impact of konjac glucomannan on hydrogels made from heat-induced pea protein isolates [26]. At low ionic strength and neutral pH, the effects of increasing konjac glucomannan polysaccharide (KGM) content on thermally induced pea protein isolate (PPI) hydrogels' thermal, textural, and rheological characteristics were explored.

2.1.2.5 Hydrophobic Interactions

Hydrogels can be produced by employing an equilibrium portion of hydrophilic and hydrophobic polymers in an amphiphilic copolymer [27]. Through hydrophobic interactions, the polymer cross-linking process may generate a wide variety of hydrogels. The interactions between the hydrophobic units in the polymer lead to the creation of hydrophobic links when it is heated and dehydrated. Because the polymer solution's heating process quickens the gelation process, this connection is simple. It is intended for these hydrophobic interactions to crosslink the hydrogels. Block copolymers that are amphiphilic include two-block, three-block, and multi-segment copolymers. The properties of the amphipaths block copolymer were significantly impacted by the kind of polymer and its concentration in solution, as hydrogel forms at an ideal concentration of the polymeric solution [28].

2.1.3 The Grafting Process

This method involves grafting, which is the polymerization of a dimer on a prepared polymer chain or placing a monomer covalently attached to another polymer host particle. This approach uses chemical reagents or high-intensity radiation to activate the polymer chain. Branching and eventually conjugating the monomers with the macromolecules arise from the production of functional monomers on stimulated macroradicals [29].

2.1.3.1 The Process of Chemical Grafting

A chemical reactant stimulates the macromolecular backbones during the chemical grafting process. Hydrogels' strong adherence to solids is crucial for their steady operation in a variety of real-world applications. However, low interfacial bonding along with solid surfaces is a problem for current hydrogel adhesion. Hao et al. [30] suggested a generic super wetting-assisted interfacial polymerization (SAIP) method to create more dense interfacial covalent bonds, which would stably attach hydrogels onto solid surfaces. The approach relied on the initiator coming into complete contact with solid surfaces by superwetting, which increases the interfacial grafting efficiency. Strong bulk failure, with a high breaking strength of $\approx$ 1.37 MPa, was demonstrated by the developed anchored hydrogels, in contrast to mild interfacial failure found in conventional techniques. The resistance of hydrogels to swelling degradation was significantly improved by the high interfacial adhesion. This work provided fresh ideas for creating hydrogels with strong anchors.

2.1.3.2 Grafting with Radiation

High-energy gamma radiation or electron beams can initiate grafting. Ortega et al. [31] investigated the process of creating a thermo- and pH-responsive binary graft using ionizing radiation on chitosan gels (net-CS). The graft is founded on pH-sensitive acrylic acid (AAc) monomers and N-vinyl caprolactam (NVCL). The direct approach produced the highest percentages (~ 40%) at low doses (10–12 kGy) and was the best synthesis radiation method. This resulted in a material with a critical pH value of 3.8 and an upper critical solution temperature (UCST) of 33.5 °C, demonstrating the system's greater hydrophilicity at low pH levels and higher temperatures. Diclofenac was used in investigations on load and release. Compared to net-CS, which released just 80% of the diclofenac after 200 min, the grafted system (32%) was able to load 19.3 mg g^{-1} and release around 95% in the same time frame. The grafted system laded 27.6 mg g^{-1} of diclofenac when it was protonated beforehand. However, electrostatic interactions caused the medication to be tightly held in the material, releasing just around 20% of it.

2.1.4 Crosslinking via Enzyme

This is a new technique for creating in situ hydrogels that involves the crosslinking of polymeric chains by enzymes. Phosphopantetheine transferase, tyrosinase lysyl oxidase, phosphatases, and plasma amine oxidase were included in the enzymes employed for the hydrogel creation. Ag-carboxamide grouping of a protein or peptide-linkage glutamine and a free amine grouping of a protein or peptide-linkage lysine can form very resistant covalent bonds due to TG, a thiol enzyme [32]. The aim of the work done by Kolour et al. [33] was to create a non-toxic wound dressing with just two natural polymers and an enzyme. A 10% weight-based microbial transglutaminase addition enhanced the dressing's mechanical properties. The rate of swelling was enhanced by 70%, the rate of deswelling by 15%, the rate of conversion by 9%, and the crosslinking density by 20% using this formulation. Furthermore, the non-toxic dressing displayed a 106% cell viability rate. Burst release behavior was seen in drug delivery testing, which is good for open wounds.

The primary cause of mortality in cases of surgical trauma and unintentional accidents is uncontrollable bleeding. The elevated pressure and speed of arterial blood flow make it difficult to stop the bleeding, resulting in an arterial rupture. Because of their excellent hemostatic effectiveness and specificity, enzyme-responsive hydrogels have become more attractive as options for hemostasis following arterial rupture. Nevertheless, the poor crosslinking of the current enzyme-responsive hydrogels remains a problem. The research studies conducted by Sun et al. [34] described an invention of a nanoporous hemostatic material for percutaneous injection that combines the Schiff-base reaction with the transglutaminase enzyme reaction. They demonstrated that the hemostatic material's ensuing burst pressure was 134 ± 10 mmHg, which was far higher compared to the average systolic blood pressure. The ability of the material to effectively halt the bleeding in the pig and rat arterial framework was noteworthy. It gelled in about 10 s and eliminated the requirement for hemostatic clamps at both ends of the vascular lesion. Additionally, in a rat liver model, the hemostatic substance served as an anti-adhesion barrier. The work offered an efficient injectable hydrogel solution that may be used to treat bleeding disorders in parenchymal organs quickly and with minimally invasive methods.

2.1.4.1 Crosslinking by Radiation

Since radiation crosslinking preserves a polymer's biocompatibility without the need for chemical additions, it is a widely used method. In addition, it is possible to change and sterilize biopolymers in a single process, making it an economical method for changing them for a specific purpose in biomedical utilizations [35]. The method is mostly dependent relating to the polymer-producing radical species as a result of being exposed to strong energy sources like gamma, X, and electron beams. The radiation impact (direct or indirect) will depend on the polymer phase, which includes rigid-state, diluted solution, and highly concentrated solution [36].

A brief illustration of the merits and demerits of the several crosslinking approaches is given in Table 2.1.

2.2 Conversion of Hydrogels to Ferrogels

The typical components of magnetic hydrogels are a polymer matrix and magnetic nanoparticles embedded inside the matrix. Numerous factors, including the kind of hydrogel and magnetic nanoparticles used, their concentration, and the size and

Table 2.1 Merits and demerits of various crosslinking processes of hydrogels [37]

Crosslinkage		Merits	Demerits
Physical	Self-formation	Reversible system, harmony with biological systems, shearing and self-repairing	Prolonged self-healing, inadequate mechanical qualities, and more post-crosslinking
	Ionic interactions	Reversible system, compatibility with biological systems, working in a moderate environment	Additional post-crosslinking, inadequate mechanical qualities, complete in terms of the ions quantity
	Heat-induced crosslinking	Reversible system, compatibility with biological systems and quick reconstruction into the hydrogel, operates in a physiological setting	Additional post-crosslinking and the ideal temperature for cell survival
	Stereo-complexation	Reversible mechanism, compatibility with biological systems	Additional post-crosslinking, poor mechanical properties
Chemical	Photo-induced	Fast gelation, room temperature settings, spatiotemporal control of the process, stabilization of weakly crosslinked hydrogels	Cells may be impacted by light irradiation, accurate identification of the photo-initiator, light intensity, and exposure duration are important
	Click chemistry Diels–Alder Strain-promoted-Azide alkyne cycloaddition Thiol-ene Oxime Thiol-Michael Aldehyde-hydrazide	Quick gel formation, gentle circumstances, impulsive response, favorable mechanical qualities, not water- or oxygen-sensitive (Thiol-ene)	Extended gelation in the absence of an initiator (Diels–Alder synthesis), several stages in the synthesis of cyclooctynes (SP-AAC), using Thiol-ene as an initiator, cells may be affected by basic pH (Oxime)
Enzymatic		Lack of exogenous reagents, spontaneous reaction, reaction control, specificity, quick gelation, and gentle circumstances	Requires additional catalyst (enzyme): the activity of the stock solution may alter while it is being stored. The expense of the enzyme is extra

distribution of the magnetic nanoparticles within the hydrogels, affect the properties of magnetic hydrogels, including their magnetic response. Several methods, such as mixing, in situ precipitation, and grafting-onto procedures, have been developed to produce magnetic hydrogels, as seen in Fig. 2.2 and Table 2.2 [38]. Since magnetic

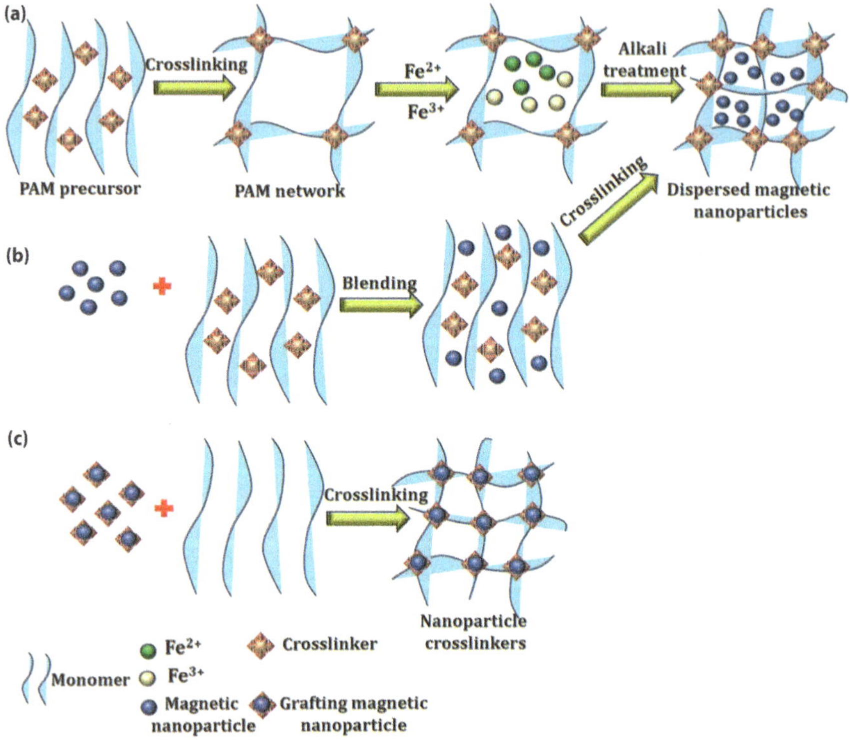

Fig. 2.2 Diagrammatic representation of the design process for magnetic hydrogel. **A** In situ precipitation method: an in situ precipitation process occurs within the network of polymer hydrogels after the crosslinking reaction, producing magnetic nanoparticles. **B** Blending procedure entails mixing a hydrogel precursor solution with the generated magnetic nanoparticles at a certain molar ratio, followed by crosslinking the hydrogels to trap the MNPs. **C** By grafting several functional groups as nano-crosslinkers onto the surface of magnetic nanoparticles, the grafting-onto technique produces the crosslinking process. This concept was adapted from Refs. [39, 40] and redrawn by the authors

PAM ferrogels are the primary topic of this chapter, PAM ferrogels are taken into account while discussing the approaches that follow.

2.2.1 Direct Precipitation Method

When using the direct precipitation approach, an iron-ion-containing inorganic salt, swollen in hydrogel interacts with precipitating reagents (such as NaOH or $NH_3 \cdot H_2O$) in the hydrogel networks functioning as a chemical reactor to produce magnetic nanoparticles (MNPs) [55]. Figure 2.2A illustrates a typical in situ precipitation technique approach. First, temperature changes, radical polymerization, or photo-crosslinking reactions are used to create hydrogels. Once swelling equilibrium is

Table 2.2 Different methods for creating magneto-responsive ferrogels

Design strategy	Hydrogel composition	Magnetic nanoparticle	Primary application	References
In situ precipitation	Chitosan-based hydrogel	Fe_3O_4	Controlled drug release	[41, 42]
	Alginate hydrogel	Fe_3O_4	General biomedical usage	[43, 44]
	Starch-grafted poly(acrylic acid) with cellulose	Fe_3O_4	Heavy metal ion removal from aqueous environments	[45]
	Polyacrylamide (PAM) hydrogel	Nickel (Ni)	Biosensing applications	[46]
Blending method	Collagen and chitosan composite hydrogel	Fe_3O_4	Biomedical applications	[47]
	Copolymer hydrogel (poly(styrene–NIPAM–MMA–PEGMA))	SiO_2/Fe_3O_4	MRI, thermal therapy, and biosensing	[48]
	Chitosan-based hydrogel	γ-Fe_2O_3	Biomedical applications	[49]
	Polyacrylamide and sodium alginate composite	Fe_3O_4	Biomedical applications	[50]
	Chitosan/PVA/laponite-based hydrogel	Fe_3O_4	Biomedical use	[51]
Grafting-onto method	Poly(acrylic acid) (PAA) hydrogel	$CoFe_2O_4$	Support for dental pulp stem cell activity	[52]
	PAM-based hydrogel	$CoFe_2O_4$	Biomedical applications	[53]
	Carboxymethylcellulose hydrogel	$CoFe_2O_4$	Drug delivery platforms	[54]

attained, the hydrogels are submerged in a concentrated aqueous solution rich in Fe^{2+} and Fe^{3+}, and ferrous ions are added in a 1:2 molar ratio. The enlarged hydrogels that have taken up Fe^{2+} and Fe^{3+} are then submerged in an alkali solution to precipitate magnetic nanoparticles. The following reaction dictates how MNPs are manufactured in the hydrogel network:

$$Fe^{2+} + 2Fe^{3+} + 8OH^- \rightarrow Fe_3O_4(\downarrow) + 4H_2O. \tag{2.1}$$

The iron salts, alkali concentrations, and the inflated hydrogel matrix's characteristics affect the reaction's yield in the hydrogel phase [56]. By precipitating magnetic nanoparticles in situ inside polyacrylamide and gum acacia (PAAm-GA) composite hydrogels, Nagireddy et al. [57] created magnetic hydrogels with uniformly dispersed

MNPs within hydrogels. Recently, Mei et al. created a class of ultra-tough polyacrylamide-chitosan-based Fe_3O_4/chitosan-polyacrylamide magnetic hydrogels with conductivity. Under an alternating magnetic field, the magnetic functionalized hydrogels could produce a significant heat content, indicating possible uses in medication release and tissue heating. Applications using magnetic heating have made extensive use of paramagnetic materials. The hydrogel samples with varying nanoparticle contents were subjected to an alternating magnetic field (AMF) to explore the magnetic-heating capabilities (Fig. 2.3a). The thermal images (Fig. 2.3b) obtained with an IR camera demonstrate that, in just two minutes with 1.5 kW, the samples' final temperatures increased from around 18 to 36.29, 48.19, and 79.26 °C (Fig. 2.3c). This suggests that the Fe_3O_4 NPs provided polyacrylamide-based ferrogels with good magnetic-heating characteristics. Furthermore, it was noticed that when the external temperature was around 18 °C, the samples under 1.0, 1.5, and 2.0 kW reached ultimate temperatures of 41.92 °C, 79.26 °C, and 98.96 °C, respectively, within 2 min of 0.6 g/mL nanoparticle (Fig. 2.3d). Magnetic polyacrylamide-based ferrogels may have applicability in tissue hyperthermia since they also produce a significant quantity of heat when subjected to alternating magnetic field, like Fe_3O_4 nanoparticles, which have already been employed as a successful therapeutic technique in the treatment of hyperthermia for malignancies. Fe_3O_4/CS-PAAm hydrogels were investigated for their potential in tissue hyperthermia using an AMF that examined a double-layer substrate consisting of gel with a thickness of 2 and 3 mm skin tissue (Fig. 2.3e). Under 2 and 3 kW of AMF power, the tissue temperature increased rapidly from 17.83 °C to 37.16 °C and 18.15 °C to 44.60 °C in 3 min, respectively. Given that cancer cells cannot survive at temperatures higher than 41 °C, this finding indicates that Fe_3O_4/chitosan-polyacrylamide gels have strong ability to tissue heat which might be useful in hyperthermic treatment of tissue. Furthermore, magnetic fields might be used to readily adjust the deformations of the gels, enabling them to be used as soft magnetic response actuators. The gels were also highly conductive and sensitive to strain, which made them suitable options for flexible electronic sensors.

Using magnetite and poly(acrylic acid-co-acrylamide)/polyacrylamide-based pseudo-interpenetrating polymer networks (pIPNs), new composite hydrogels were created through precipitating Fe^{3+}/Fe^{2+} ions inside the hydrogel structure by Simeonov et al. [59]. The crystallinity of the magnetite particles increased in proportion to the polyacrylamide concentration in the pIPN ingredient. Differential scanning calorimetry analysis of the composites' thermal characteristics reveals a rise in the glass transition temperature of the resultant composites is influenced by the poly (acryl acid)/polyacrylic amide copolymer ratio in the components of pIPNs. In addition, the composite hydrogels show superparamagnetic and responsiveness to ionic strength and pH. The research demonstrated the feasibility of using pIPNs as a matrix for regulated inorganic particle deposition in the synthesis of polymer nanocomposites.

The direct precipitation method provides several advantages when developing magnetic hydrogels. Initially, a substantial number of inorganic particulates can be incorporated into the hydrogel network, while ensuring the uniform dispersion of colloidal particles across the matrix. Secondly, it is a cheap and easy process to

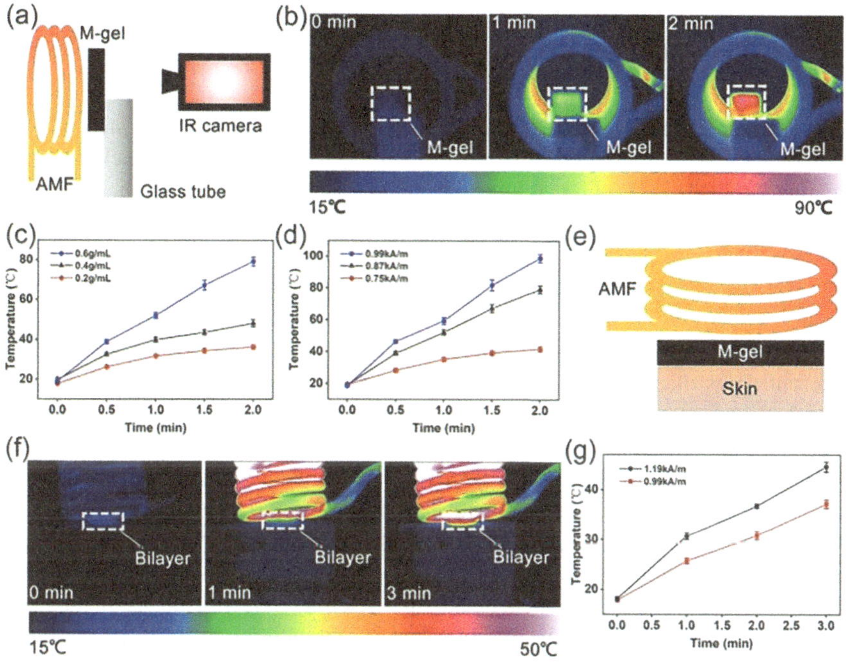

Fig. 2.3 Fe_3O_4/CS-PAAm DN hydrogels' magnetic-heating characteristics and biological uses. **a** Schematic representation of the magnetized hydrogel sheet heating tests under the influence of an alternating magnetic field (AMF). **b** Thermal pictures of the hydrogel film at 0, 1, and 2 min (0.6 g/mL Nanoparticle, 1.5 kW). **c** Gel sheets' temperature was at 1.5 kw (0.87 kA/m) with a frequency of 350 kHz at various times when 0.2, 0.4, and 0.6 g/mL nanoparticles were added. **d** Gel sheets with 0.6 g/mL nanoparticles at various times measured at 0.75, 0.87, and 0.99 kA/m (power = 1.0, 1.5, and 2.0 kW, respectively) with frequency of 350 kHz. The heating tests are conducted at a temperature of 15 °C. **e** Representative illustration of hydrogel-skin bilayer thermal studies in an AMF. **f** Hydrogel-skin bilayer thermal imaging at 0, 1, and 3 min (0.6 g/mL nanoparticle, 3 kW). **g** Thermal condition of bilayers of hydrogel and outer surface at various times under 0.99 and 1.19 kA/m (applied power of 2.0 and 2.5 kW, respectively) with a frequency of 350 kHz. The hydrogels used in the bilayer heating studies have a nanoparticle concentration of 0.6 g/mL, with the heating treatment outside being 13 °C. Adapted with permission from Ref. [58]. Copyright 2023 Elsevier

prepare. The hydrogel matrix may be compromised by an alkaline solution during the formation procedure, so the direct precipitation method is only appropriate for certain hydrogels with a reliable matrix [60]. Additionally, the resultant Fe_3O_4 magnetic nanoparticles may be poor in hydrogels, which include negatively charged functional groups (such as carboxylic groups), as these groups can interact with iron Fe^{2+} and Fe^{3+} in salt solution to create coordination compounds.

2.2.2 Using Blending to Create PAM Ferrogels that Respond to Magnetism

Magnetic nanoparticles and hydrogels are made individually and in a step-by-step manner for the mixing procedure (Fig. 2.2B). The most common technique for creating MNPs (like Fe_3O_4) is the traditional coprecipitation approach. The MNP sediment that has been produced is subsequently distributed in an oily or aqueous phase, or ferrofluid, to prevent oxidation and grouping. The MNPs are encapsulated in the hydrogel when the combination of the ferrofluid and the hydrogel precursor solution undergoes crosslinking. Zhang et al. [50] used a simple blending method to design PAAm-SAlg (polyacrylamide and sodium alginate) organization. Two different types of chemical bonds were used to build the crosslinkage connection in such a magnetic hydrogel. Physical and chemical bonding interactions enshrine magnetic nanoparticles with Fe_3O_4 cores in hydrogel networks. Hydrogel properties are affected differently when magnetic particles are added. First, the stretching resistance of magnetic hydrogels rises by 40% from 0.035 to 0.049 MPa when contrasted with hydrogels without magnetic nanoparticles; on the other hand, the expanding property is marginally diminished. Additionally, the printed fiber of hydrogel has a diameter of 100 μm with the magnetic nanoparticles uniformly dispersed in the fiber; additionally, improving the air pressure during fabricating will result in randomly distributed lumps on the hydrogel surface; and finally, magnetic hydrogels maintain the shear-thinning behavior and exceptional superparamagnetism, though the magnetic response is slightly weaker than genuine magnetic nanoparticles.

Ahmad and the research team [48] simply blended iron (II, III) oxide/silicon dioxide nanocomposite particles and stimulus-sensitive hydrogel nanoparticles to create two distinct magnetic and stimulus-sensitive composite hydrogel nanoparticles with varying external properties for this experiment. The hydrogel particles produced exhibited temperature sensitivity in poly(styrene-N-isopropylacry-lamide-methyl methacrylate-polyethylene glycol methacrylate), also known as P(S-NIPAM-MMA-PEGMA), and pH and heat responsiveness in poly(S-NIPAM-methacrylic acid-PEGMA), also known as P(S-NIPAM-MAA-PEGMA). Both magnetic and stimulus-sensitive iron (II, III) oxide/silicon dioxide /P(S-NIPAM-MMA-PEGMA) and iron (II, III) oxide/silicon dioxide/P(S-NIPAM-MAA-PEGMA) composite hydrogel nanoparticles exhibited temperature-responsive absorptions of biomolecules. The blending approach allows for the control of reactant content, stirring rate, and preparation process that are optimized to produce magnetic nanoparticles with uniform particle size in hydrogels across a wide range from nanometers up to micrometers. Moreover, since magnetic nanoparticle encapsulation and preparation are carried out separately, the procedure is easy to manage. In contrast, obtaining an evenly magnetic nanoparticle dispersion within the hydrogels is challenging, and when placed in a liquid solution, magnetic nanoparticles may leach out of the magnetic hydrogels.

2.2.3 Grafting-Onto Process

The magnetic nanoparticles and the hydrogel matrix do not form covalent bonds in either the blending technique or in situ precipitation. Therefore, it is impossible to ensure the reliability of the magnetic nanoparticles scattered throughout the hydrogels [48]. By adding multiple functionalized groups to the outer surface of the magnetic nanoparticles, the grafting-onto method creates a covalent linkage between the hydrogel matrix and the magnetic nanoparticles. These MNPs then function as nanostructured-crosslinkers, forming a covalent bonding with the monomers when polymerized, as shown in Fig. 2.2C. Magnetic $CoFe_2O_4$ NPs were reported to be incorporated into polyacrylamide (PAAm) hydrogel networks by Schmidt and colleagues [53]. During the polymerization process, Cobalt (II) Ferrite nanoparticles were surface-modified with unsaturated methacrylic groups using a siloxane approach. These groups were able to create a covalent coupling with the hydrogel matrix based on PAAm. The direct chemical interaction between the hydrogel network and the MNPs ensures that they remain confined within the hydrogel without leakage, which is a key feature of the grafting-onto method (Table 2.2). However, the method's wide range of applications is restricted by its intricate creation process, protracted preparation time, and high-related costs.

Both the magnetic component (embedded MNPs) and the constituent hydrogel network affect the mechanical, biocompatible, biodegradable, and magnetic response characteristics of the magnetic hydrogel system. Magnetic-responsive hydrogels can be created using natural or synthetic polymers. As scaffolding for cell development and adhesion, natural sources of polymers, namely proteins (such as gelatin, collagen, and fibrin) as well as polysaccharides (such as dextran, chitosan, and alginate) have vital physiological roles (see Table 2.2). However, because natural polymers do not include MNP active sites, they might avoid appropriate for creating magnetic-responsive hydrogels using grafted magnetic nanoparticles. Their poor mechanical qualities are another drawback for natural polymer-based magnetic hydrogels, which might limit their application in areas such as tissue engineering. Magnetic hydrogels, conversely, fabricated from synthetic polymers (PAM) allow for the methodical regulation of chemical formulation, which may be advantageous for the grafting of MNPs onto hydrogel networks, the direct precipitation of magnetic nanoparticles in hydrogels, and the control of mechanical and biodegradable properties [61–64]. Furthermore, the magnetic-responsive behavior of magnetic hydrogels is significantly influenced by magnetic composites, such as MNPs. The concentration and magnetization at the saturation of magnetic nanoparticles determine the magnetization of magnetic hydrogels, and these may be indicated by using the succeeding equation.

$$M = \varphi_m M_s \left(\coth \varepsilon - \frac{1}{\varepsilon} \right), \tag{2.2}$$

where M_s is the saturation magnetization, φ_m is the volume percentage of magnetic nanoparticles in the hydrogel, and ε is defined as $\varepsilon = mH/k_B T$. The variables m is the gigantic magnetic moment of the magnetic nanoparticles, H is an extrinsic magnetic field, k_B stands for Boltzmann constant, and T shows temperature [65].

Because the component polymeric networks of hydrogels and ferrogels (HG-FGs) include hydrophilic groups, the 3D arrangements of polymers can both retain and hold water. The word "network" suggests that HG-FGs have crosslinks, which give them pure elastic to viscoelastic characteristics. HG-FGs are usually prepared using hydrophilic monomers; however, to tailor the chemical, mechanical, and physical characteristics for particular uses, hydrophobic monomers are often included. Polymeric networks, both natural and artificial, can be thought of as HG-FGs. Chemically, synthetic polymers are often stronger and less hydrophobic than natural polymers. When combined with natural polymers, they offer a mechanically strong structure and a moderate rate of breakdown [66–68]. HG-FGs are formed by the processes of crosslinking and polymerization, which establish strong bonds between polymers. The way that polymerization and crosslink differ is that in the former case, a catalyst is often used to help bond a polymer and its monomers. On the other hand, a crosslink is a sequence of covalent or ionic connections that connects two neighboring polymer chains. Thus, after a thorough discussion on crosslinking processes, the polymerization approaches are deliberated in a subsequent section.

2.3 Typical Polymerization Approaches for Hydrogels and Ferrogels

Mixing monomers that react differently to diverse stimuli in the environment is the most practical way to create HG-FGs. HG-FGs can be made using any method which is suitable for creating a network of crosslinked polymers. Hydrogels are often created by combining multifunctional crosslinkers with natural or synthetic hydrophilic monomers by free radical polymerization processes [69]. The characteristics of the resulting HG-FGs are significantly influenced by the choice of polymerization process/technique [70]. HG-FGs can be made step-by-step by generating polymers having active groups that are capable of crosslinking each other or reacting with suitable linking agents, or they can be made in one step by polymerizing and crosslinking multifunctional monomers simultaneously [71, 72]. To get the optimum synergistic impact of those materials for applications, various types of inorganic, organic, and polymeric components are combined to create composite HG-FGs. For example, the incorporation of hydrophobic polymer fragments in the existence of a crosslinking agent might enable the material to adhere to a certain shape and attain the appropriate mechanical strength [73].

2.3.1 Chain-Growth Polymerization of HG-FGs

To create chemically crosslinked HG-FGs, chain-growth polymerization using a free radical process is widely used. Three phases make up the polymerization process: start, propagation, and termination. Hydrophilic in most cases, –C=C– containing monomers take part in polymerization via free radicals [74]. The following describes many methods of free radical polymerization for the creation of HG-FGs.

Because bulk polymerization (Fig. 2.4a) is a straightforward process, it is frequently used to create HG-FGs. Using a limited number of crosslinkers, this approach polymerizes liquid monomers and initiators soluble in monomers. Chemical catalysts, radiation, and/or UV light are widely employed to initiate the polymerization process. Bulk polymerization features a faster rate of polymerization, ineffective heat control, and a sharp rise in reaction viscosity. To fine-tune the HG-FG's characteristics, it is imperative to regulate the pace of conversion. When submerged in water, the resulting HG-FGs' glassy, clear polymer network expands and gains flexibility [75]. The bulk polymerization conversion rate may be regulated by varying the thermal state and activator concentration. Other strategies would be to limit the reaction at a low transformation; however, that is thought to be unfeasible in large-scale operations. Therefore, various methods of polymerization such as emulsion, suspension, and solution polymerization are extensively used in the production of HG-FGs [76].

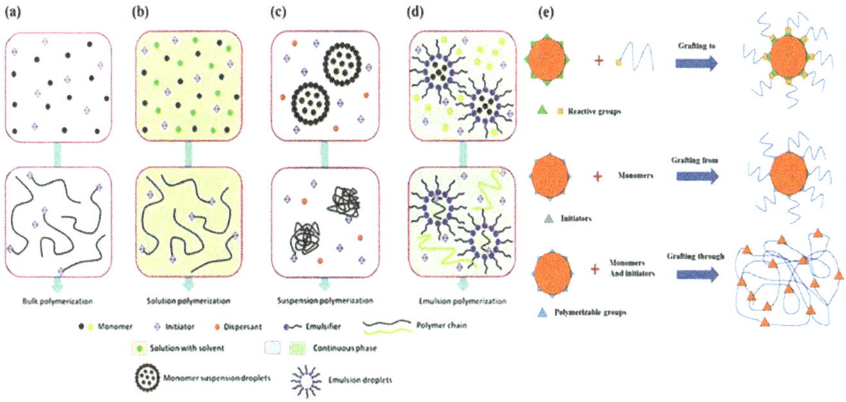

Fig. 2.4 Shows a schematic diagram of popular methods for free-radical polymerization. **a** Polymerization in bulk. **b** Polymerization in solution. **c** Polymerization by suspension. **d** Emulsion polymerization. **e** Graft polymerization methods. Reproduced with permission from Ref. [77]. Copyright 2021 Royal Society of Chemistry (Open Access under the license CC-BY-NC-ND)

2.3.1.1 Polymerization in Solution

In solution polymer formation, multifunctional crosslinkers and ionized or non-ionized monomers are combined with solubilizers (such as benzyl alcohol, water-ethanol or water, ethanol, mixes) and initiation via ultra-violent or redox reactions to produce hydrogels. After the hydrogel is isolated, distilled water is used to remove any residual monomers, crosslinking agents, activators, and other impurities (Fig. 2.4b) [78]. Solution polymerization is widely recognized for its simplicity in synthesis, cheap processing costs, and improved control over heat management during the polymerization process. Polymerization is harmless with innocuous since it occurs in a water-rich medium [74]. To create superabsorbent hydrogels based on cellulose, solution polymerization is frequently employed. Room temperature conditions are suitable for reaction and have a high rate of polymerization. Solution polymerization produces superior heat transmission and dissipation than bulk polymerization because the solution's low viscosity makes it easier to agitate the reaction mixture [79]. Both bulk and solution polymerization processes are typically homogeneous, yet they can turn heterogeneous if the polymer that is produced is insoluble in the solvent and monomer used in the corresponding process [79].

2.3.1.2 Polymerization in Suspension, Encompassing Inverse-Suspension Polymerization

In general, insoluble monomers and initiators combiner with suspending agent that has low hydrophilic-lipophilic balance in a hydrous solution are used in suspension polymerization (Fig. 2.4c). They are continuously stirred to produce 0.1–5 mm width droplets of monomers. The polymeric hydrogel particles are created as the polymerization process moves on and may be filtered out of the reaction fluid. Here, bulk polymerization on a tiny scale is carried out by individual monomers. Water is a great medium for heat transmission since it is the normal setting. To protect the droplets from blending, a protective dispersing agent such as polyvinyl alcohol (PVA) and methylcellulose (MC) or carboxymethyl cellulose (CMC) is frequently used [79]. For the creation of hydrogels, inverse-suspension polymerization is commonly employed [80].

2.3.1.3 Emulsion Polymerization or Micellar Polymerization

Micellar polymerization may also be employed to synthesize hydrogels. A hydrophilic initiator, a surface-active agent, a crosslinking agent, and minuscule water-soluble monomers that is, marginally water-soluble or entirely hydrophobic monomers are all involved in an emulsion polymerization process [81, 82]. Inverse emulsion polymerization employs an organic liquid's hydrophilic monomer. A typical method of emulsion polymerization is shown in Fig. 2.4d [78]. Compared to suspension polymerization, it yields significantly smaller polymer particles (0.1–3 μm). Heat transfer over bulk polymerization may be accomplished efficiently, and the suspension and emulsion polymerization processes are simply regulated [79]. Emulsion polymerization employs an initiator that is insoluble in water, as opposed to suspension polymerization.

2.3.1.4 Mechanism of Graft Polymerization

When employing a grafting method, the bulk polymerized hydrogel demonstrates poor mechanical characteristics that can be generated, individually when grafted onto more robust support grids. On the surface of the support, free radical sites are often produced where the polymerization of monomers occurs directly to create enhanced covalent connections accompanied by the support grid. For instance, it's typical practice to graft vinyl monomers onto polysaccharides [31]. Furthermore, pH-responsive hydrogels made from grafted poly (acrylic acid) from hybridized chitosan (CHT) along with cellulose through thiourea formaldehyde resin are mechanically stronger than hydrogels made from grafted AAc from CHT [83]. Figure 2.4e shows an alternative grafting method for hydrogel synthesis.

2.3.2 Step-Growth Polymerization of HG-FGs

Hydrogels are created via condensation polymerization, which utilizes different functionalized groups that include monomers. The condensation polymerization method is propagated by the reaction and formation of covalent bonds between these complementary functional groups [74]. The mechanical properties of photo-responsive hydrogels produced by the chain and condensation polymerization process were discovered by Tibbitt et al. [84] (Fig. 2.5). The results showed that the step-growth hydrogels' mechanical stability, tensile strength, ductility, and shear deformation to yield were superior to those of the chain-growth hydrogels due to the cooperativity and homogeneity of the network. However, because chain-grown hydrogels had a better network connection, the erosion rate due to light treatment was lower for these hydrogels. It is possible to give hydrogels the desired qualities by comprehending hydrogel networking.

2.4 Physical Properties of Hydrogels and Ferrogels

To create HG-FGs for biomedical implementation, it is highly beneficial to utilize natural or artificial polymers that have hydrophilic substituent groups. These groups help in absorbing large amounts of water and also promote interaction with biological tissues like mucous membranes and epithelial layers. When HG-FGs are completely

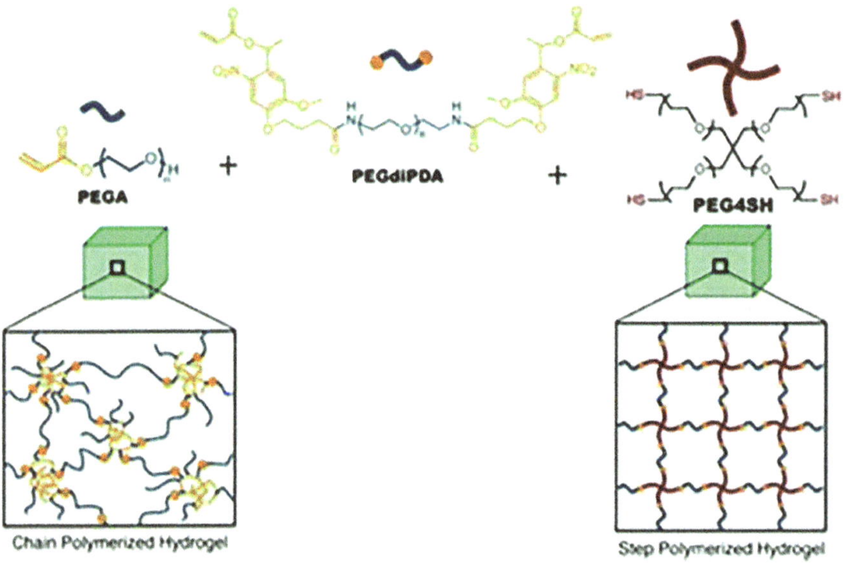

Fig. 2.5 Photolabile monomer (PEGdiDPA) is polymerized using a chain and step process to create a photodegradable hydrogel. **a** Chain polymerization of PEGA and PEGdiPDA by the use of free radicals (heterogeneous network structure, for example). **b** Michael addition polymerization step-polymerization of PEGdiDPA at the ends with PEG4SH. *Thiol-functionalized tetra-arm poly (ethylene glycol) (PEG4SH); PEGA (monoarylated poly(ethylene glycol)); poly(ethylene glycol) di-photodegradable acrylate (PEGdiPDA). Reproduced with permission from Ref. [77]. Copyright 2021 Royal Society of Chemistry (Open Access under the license CC-BY-NC-ND)

swelled, they often have poor interfacial angle with physiological fluids, and are soft, flexible, and nearly viscoelastic, which minimizes the adverse immunological response. These elements are all involved in HG-FGs' biocompatibility. Additionally, the HG-FGs often degrade to varying degrees based on the kind of crosslinker used.

Moreover, the most significant feature of HG-FGs is their capability to swell. Three processes occur when HG-FGs swell: (i) water diffuses into the network of HG-FG molecules (known as primary bound water); (ii) polymer backbones relax; (more water entering the network is known as secondary bound water); (iii) the HG-FG matrix swells (more water entering the network is referred to as free water). The elasticity character of the polymer strands and their ability for water molecules are the factors that cause swelling, following the Flory–Reihner theory [85]. When environmental stimuli vary, HG-FGs respond differently. These stimuli may be broadly classified as (i) physical (light, temperature, etc.), (ii) chemical (ionic strength and pH), and (iii) physiological (enzymes). Apart from temperature, that may be either internal or exterior, chemical and physiological activators are internal, whereas physical activators are external. In addition to all of this, shape memory HG-FGs are a unique kind of activator-responsive smart HG-FGs that has dual features: (i) a permanent pattern and (ii) a chemical or physical signaling that can aid in restoring its original form [86].

HG-FGs with ionic pendant groups that are covalently crosslinked and exhibit pH responsiveness are known as pH-responsive HG-FGs. Because of electrostatic repulsion caused by pendant group ionization, HG-FGs containing anionic groups swell at higher pH values. On the other hand, hydrogels with cationic groups swell at low pH values. Temperature-sensitive HG-FGs are those that exhibit swelling or deswelling behavior in response to thermal variations as a result of amendments in the interactions between polymers or water and polymers. One kind of HG-FG that is sensitive to temperature is poly (N-isopropylacrylamide) (PNIPAM) hydrogel [87]. HG-FGs can exhibit expansion at a temperature more than the upper critical solution temperature (UCST) [88], swelling at a temperature lesser than the lower critical solution temperature (LCST) [negatively temperature-sensitive], or thermally reversible hydrogels that exhibit sol-gel transition both upward and downward with an aspect temperature. Thermally reversible HG-FGs are referred to as poloxamer HG-FGs [89].

2.4.1 Swelling Kinetics of IIG-FGs

HG-FG's capability to enlarge in many network settings, involving pH, water, and ionic strength, makes it beneficial in various implementations. The HG-FG interacts with the biological and atmospheric conditions in its environment, such as temperature, ionic medium, pH solvent, and electric field, with radiant exposure [90, 91]. The swelling kinetics and equilibrium are affected by several factors, such as the synthesis conditions, ionic environment, crosslinking degree, and chemical composition of the polymers. The swelling ratio is defined as the weight ratio of the swollen gel to the dry gel, and it is used to quantify the extent of swelling. The swelling ratio of a gel is affected by the degree of crosslinking; gels with higher crosslinking exhibit lower swelling ratios, and vice versa. The swelling behavior is greatly impacted by the chemical structure due to the presence of hydrophobic and hydrophobic groups on the polymeric chains. In contrast to hydrophobic groups, HG-FGs with greater hydrophilic groups swell more. HG-FG swelling is also affected by temperature and pH. Due to the ionization of hydrophilic groups with variation in pH, pH-responsive HG-FGs swell. Ionization interferes with the secondary interactions between polymer chains and leads to electrostatic repulsion between similar charges on the polymer. Three processes are involved in the swelling of HG-FGs: the infiltration of H_2O into the network, the relaxation of polymer strands, and the enlargement of the HG-FGs matrix. When HG-FGs are dried, it is referred to as the glassy state; when they absorb water, it is referred to as the rubbery state. When glassy or dry HG-FG encounters the aquatic media, the solvent molecules can occupy voids created by gaps in polymer chains. The glassy state transitions into the rubbery state called swelling when enough water has been absorbed into the HG-FG network. The HG-FG permits water to move in and out via the diffusion procedure. The morphology and swelling/ properties in salt solutions and at different pH values are shown in Fig. 2.6. In a similar vein, Bashir et al. synthesized karaya gum grafting with poly (acrylic acid)

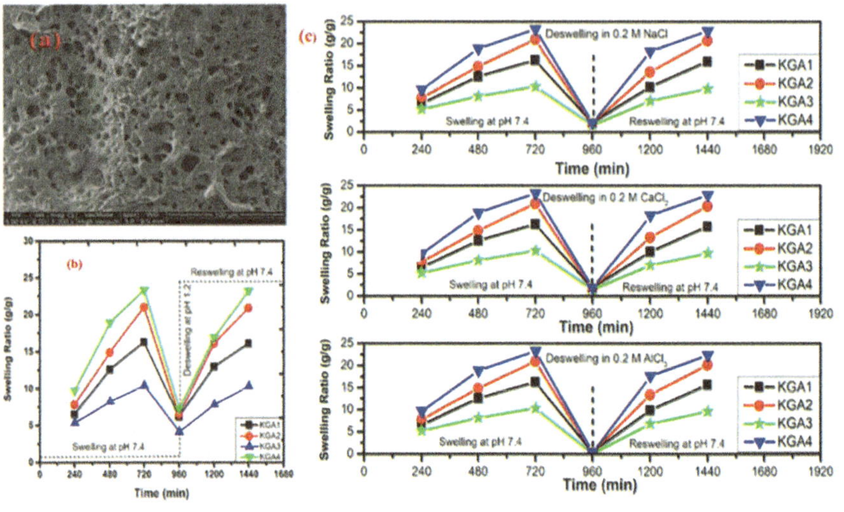

Fig. 2.6 **a** Morphology of the hydrogel karaya gum grafting poly (acrylic acid), **b** swelling-deswelling-reswelling at different pH values, **c** salt solutions. Reproduced with permission from Ref. [92]. Copyright 2020 MDPI (Open Access under the license CC-BY-NC-ND)

and studied the swelling-deswelling-reswelling of hydrogels in salt solutions and at different pH levels. Furthermore, water retention was noted as well; Fig. 2.6; [92] displays the findings.

Awasthi et al. [23] observed the samples' swelling ratio by the weight of hydrogels following their immersion in phosphate buffer saline (PBS) (Fig. 2.7a). Particle size, specific surface area, and surface and molecular structure are some of the variables that affect the rate of water absorption. After ten hours of dipping, the polyacrylamide (PAM) and hybrid hydrogel composed of PAM showed a stable expansion (Fig. 2.7a). One important feature of PAM–TiO_2–CNT hydrogel for cartilage regenerating implants is its steady volume even after 25 h, indicating the material's non-swellable behavior [43]. The decreased swelling ratio in composite hydrogels can be attributed to the stronger covalent bonding between incorporated nanoparticles and PAM. When TiO_2 and CNT are added to PAM (PAM–TiO_2–CNT), a compacted matrix of PAM is created which prevents the hydrogel from expanding.

It is possible to fit the swelling ratio of various hydrogels to a Fickian model using the quadratic trend of the swelling ratio versus time graphs. M_0 is the mass of the sample at zero swelling, or $t = 0$, M_t is the mass of the substrate after enlarging for time t, and K_s is the swelling rate (in s^{-1}). Another model that suited the swelling was non-Fickian, or anomalous.

$$\text{Swelling percentage } W(t)\% = \left(\frac{M_t - M_0}{M_0}\right)\% = K_s t^{0.5} \tag{2.3}$$

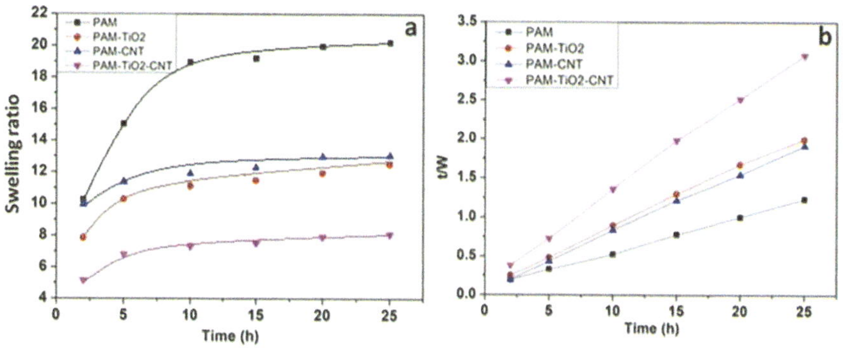

Fig. 2.7 Swelling ratio **a** pseudo-second-order dynamics of the hydrogel composites **b** of polyacrylamide, polyacrylamide-titanium dioxide polyacrylamide-carbon nanotubes, and polyacrylamide-TiO_2-carbon nanotubes are shown. Reproduced with permission from Ref. [23]. Copyright 2021 American Chemical Society

$$F = \frac{M_t}{M_\infty} - Kt^n, \tag{2.4}$$

where the equilibrium mass, measured after 25 h of swelling, is M_∞. For purely Fickian diffusion, n has a value of 0.5 and a range of 0–1. Thus, the values of K and n at different periods were obtained by utilizing $F = Kt^n$. In the ith sample, F_{2i}, F_{1i} are estimated at times t_2 and t_1, respectively, and $n_i = \frac{\log \frac{F_{2i}}{F_{1i}}}{\log \frac{t_2}{t2}}$. A measurement of n's diversion from the Fickian diffusion model was its divergence from 0.5. This variance was least in the pure PAM and highest in the PAM–CNT composite. The deviation was the same for PAM–TiO_2 and PAM–TiO_2–CNT composites.

Furthermore, a pseudo-second-order model [45] was used to explore the swelling kinetics of the genuine and matrix hydrogels. For various hydrogel-based material types, the swelling rate at any given time is a quadratic function of time as

$$\frac{dW}{dt} = K(W_e - W)^2. \tag{2.5}$$

Boundary conditions apply to this equation. As time passes from 0 to t, W changes from 0 to W_t. Upon integration, the following equation was obtained:

$$\frac{1}{W_e - W} = Kt + \frac{t}{W_e}. \tag{2.6}$$

After the Eq. (2.3) is rearranged:

$$\frac{t}{W} = \frac{t}{W_e} + \frac{1}{KW_e^{\ 2}}, \tag{2.7}$$

where the swelling ratios at time t, equilibrium, and rate constant were denoted by W, W_e and K, respectively. A straight line with a slope of $1/W_e$ and intercept of $1/(K.\ W_e^2)$ was produced by plotting t/W against t. This line may be used to calculate swelling capacity (W_e) and swelling rate constant (K), respectively. In the line equation $y = mx + C$, the plots of t/W against t are therefore linear with constants $C = \frac{1}{KW_e^2}$ and $m = \frac{1}{W_e}$. For every manufactured hydrogel, the plot between t/W and t (Fig. 2.7b) displayed a straight edge, indicating that the experimental data align well with a pseudo-second-order model. According to the results, the expanding rate constant (K) was quadrupled for the polyacrylamide-titanium dioxide-carbon nanotubes composite than for pure polyacrylamide. As opposed to pure PAM, the polyacrylamide-carbon nanotubes composite had a greater than threefold amount. It demonstrates that PAM composites (including CNT and TiO_2) require less time to obtain a swelling ratio than pure PAM. Pure PAM has the largest equilibrium swelling ratio and the shortest slope, indicating that it can store more water than its composites at equilibrium. Swelling capabilities showed theoretical values that closely match experimental results. Therefore, the pseudo-second-order model used to account for the produced hydrogels' swelling kinetics was a good fit.

Buvana et al. [93] developed composite of polyvinyl alcohol (PVA), starch, and manganese iron oxide ($MnFe_2O_4$) hybrid hydrogel membranes that are dual-stimuli-sensitive, or PS/pp-MI. By altering the concentration of $MnFe_2O_4$ NPs (pp-MI NPs) in the polymeric matrix, several hydrogels (Polystyrene, Polystyrene/Polypropylene-Manganese Oxide Composite 1, Polystyrene/Polypropylene-Manganese Oxide Composite 2, and Polystyrene/Polypropylene-Manganese Oxide Composite 3) were created. As the period spent at room temperature increases, the swelling ratio values progressively rise. It was observed that the hydrogel's ability to swell is affected by the amount of pp-MI it contains. For example, after 60 min, the plain polystyrene hydrogel displays a swelling ratio (SR) of 93.75. During the same time, the SRs of Polypropylene-Polystyrene Manganese Oxide Composite 1 (43.80), Polypropylene-Polystyrene Manganese Oxide Composite 2 (30.03), and Polypropylene-Polystyrene Manganese Oxide Composite 3 (15.35) declined. Additionally, it was noted that when the concentration of polypropylene-manganese oxide nanoparticles increased, the hydrogels' ability to swell reduced. Because of the interrelations between the implanted pp-MI nanoparticles and the polymer matrix, more crosslinking sites are formed, which explains why the swelling ratio of the hydrogels has reduced.

2.4.2 Mechanical Properties

HG-FG's tensile strength is significant for biological and pharmacological utilization. To ensure proper physiological function from multiple perspectives, such as medical uses, tendon and ligament repair, cartilage restoration, tissue engineering, wound care, and drug delivery systems, the tensile strength of the HG-FGs needs to be assessed. It is crucial to preserve the gel's structural consistency throughout the entire period of drug delivery. HG-FG can achieve adequate mechanical properties by adjusting the crosslinking level and incorporating specific polymers, co-monomers, and crosslinking agents. Increasing the crosslinking level yields a robust gel framework [94]. Conversely, too much crosslinking will result in reduced elasticity and elongation, along with heightened brittleness. Elasticity is essential for enabling the integration of medicinal components and giving the crosslinked structure increased flexibility. Therefore, HG-FGs need to achieve the optimal degree of crosslinking to achieve the balance between mechanical strength and elasticity [95].

Numerous methods may be used to evaluate the mechanical characteristics of HG-FGs, such as frequency-based evaluation utilizing rheometry and dynamic mechanical analysis, as well as compression and tensile analysis, that can be performed using either restricted or unrestricted regional grooves using a probe. When doing frequency-based sinusoidal analysis, a rheometer is typically used. Various sweep measurements can be carried out once the substrate is put into a designated geometrical equipment [96]. With the use of a rheometer, one may investigate the viscoelastic characteristics of HG-FGs. "Rheology is the study of how materials deform and flow when subjected to stress or strain." The name originates from the Greek words "rheo," meaning flow, and "-ology," which denotes study [97]. In 1929, the American Rheology Society accepted this term. For the past three hundred years, Hooke's laws and Newton's have served as mathematical descriptions of ideal liquid and solid behaviors, respectively. Nonetheless, a lot of substances exhibit a behavior that is in between that of perfect liquids and solids. Such traditional ideas cannot explain those actions. After Maxwell and Wilhelm Weber described the non-perfect behaviors of fluids and silk thread, scientists started to reconsider their views. Mathematical verification was provided for the elastic characteristics that Maxwell claimed fluids exhibited. Wilhelm Weber also investigated that silk thread had non-ideal elasticity properties. Usually, these substances were categorized as non-Newtonian and viscoelastic substances. Non-Newtonian substances' viscosity is time- and stress-dependent, but viscoelastic substances may recover from minor deformations [98]. Shear stress and strain variables are significant rheological factors. Shear stress is defined as the "ratio of the force applied per unit area that induces disarray in the materials situated between the plates." "The ratio of change in the material's displacement (x) to its height (h), which can be simply represented as tan α," is what is called strain. The internal force operating within the material controls the movement's velocity once force is applied. Rheology tests conducted at a small scale can be employed to assess the mechanical properties of hydrogel. When force is applied, the particles within a hydrogel shift about each other, leading to strain. Furthermore, it's well known that elastic materials deviate from when force is exerted, leading to elastic deformation, and then revert to their initial shape once the force is removed. Conversely, viscous materials experience irreversible strain when an external force is applied to them. The movements of the viscous materials are caused by this applied force. HG-FGs, on the other hand, have a viscoelastic quality rather than being entirely viscous or elastic. Within the linear viscoelastic region (LVR) of the substance under examination, rheological tests are conducted. The degree of the exerted stress or strain does not affect the LVR, guaranteeing the mechanical characteristics of HG-FGs. Furthermore, relatively little deformation occurs in viscoelastic materials when a gradual external force is exerted. This is caused by the polymers' still-close-to-equilibrium molecular configurations. Consequently, even for an equilibrium system, dynamic activities at the molecular level are reflected in the mechanical behavior, which occurs continuously.

Shear strain is defined for **controlled-strain rheometers** as time (t) of a sinusoidal function, that is represented by Eq. (2.8):

$$\gamma_0 = (\sin \omega t) = \gamma_{(t)}, \tag{2.8}$$

where ω is the oscillation's angular frequency and γ_0 is the amount of applied strain. It is expressed in rad/s. $\omega = 2\,\pi f$ and is the relationship between the angular frequency and the frequency measured, f, in Hertz. Moreover, sinusoidal stress is produced by sinusoidal strain, as shown by Eq. (2.9).

$$\tau_0(\sin \omega t + \delta) = \tau_{(t)}, \tag{2.9}$$

where δ is the phase difference between two waves and τ_0 is the stress amplitude.

Shear stress is specified as a sinusoidal function of time (t) for **controlled-stress rheometers**, and this may be written as in Eq. (2.10):

$$\tau_0\ (\sin\ \omega t) = \ \tau_{(t)}. \tag{2.10}$$

Equation (2.11) may be used to show how sinusoidal tension leads to sinusoidal strain.

$$\gamma_0(\sin \omega t + \delta) = \gamma_{(t)} \tag{2.11}$$

Hooke's law states that stress and strain are always in sync, and this principle applies only to elastic materials. Thus, with $\delta = 0°$, the phase angle is 0. In contrast, pure viscous materials exhibit two separate waves that have a phase angle of 90° (δ = 90°). On the other hand, the phase angle of viscoelastic materials is among 0 and 90°.

The shear Young's modulus (G'), loss modulus (G''), and loss factor, tan δ, are key parameters of hydrogel parameters in small-amplitude oscillatory studies. The deformation energy stored during a material's shearing process, which conveys its rigidity, is evaluated by Young's modulus. The energy lost during shearing, which defines the component's movement characteristics, is measured by the diminution modulus. The ratio of loss modulus to Young's modulus (loss modulus/Young's modulus) represents the loss factor, tan δ. High viscous properties are seen in materials when tan $\delta > 1$ ($G'' > G'$). Conversely, materials with tan $\delta < 1$ ($G' > G''$) possess enhanced elastic rigid characteristics. We analyze the moduli of storage and loss about strain, duration, and frequency. The observation of time-based assessment of storage and Young's modulus can be utilized to examine the gelation procedure. By monitoring these moduli of the material concerning strain, one may ascertain the LVR in which both moduli are unaffected by strain.

The moduli against frequency can be used to study an HG-FG's behavior at various time scales. The material exhibits dominating liquid behavior when $G'' > G'$ at low frequency for an extended period, and solid properties when Young's modulus >> loss modulus at high frequency for a brief period. Furthermore, two crucial variables that greatly affect the material's applicability are gelation kinetics and gel stiffness. Up to a certain amount, referred to as critical strain, the tensile characteristics of viscoelastic materials are not dependent on strain. Above this point, strain causes Young's modulus to decrease, and a nonlinear response may be seen. The beginning stage in determining a component's LVR at a frequency of one hertz (Hz) is to do strain-dependent Young's modulus and loss modulus measurements. Furthermore, on a continuous strain below the critical threshold, a material's structure may be studied. These findings add to our understanding of the role played by colloidal forces and particle interactions. G' is frequency-independent below the critical strain, demonstrating the properties of solid materials. On the other hand, the fluid-like material is represented by frequency-dependent Young's modulus. Ou et al. created a biocompatible double-network hydrogel composed of polyacrylamide, calcium alginate, and ethanol (PAM/CA/EtOH30 DN hydrogel) using solvent-exchange techniques and free radical polymerization. The preparatory method, strengthening processes, also use of polyacrylamide/calcium alginate/$EtOH_{30}$ hydrogels are shown in Fig. 2.8A. The resulting PAM/CA/$EtOH_{30}$ hydrogel demonstrated exceptional mechanical capabilities thanks to the double-network structure's energy dissipation and the synergistic impact of solvent substitution. As shown in Fig. 2.8B(a), as the ethanol concentration in a synergistic medium solution rose, the PAM/CA/EtOH hydrogel's tensile strength improved, and the volume swelling ratio decreased. Figure 2.8B(b) displays the mechanical stress-strain curves for each of the triad hydrogels. Furthermore, the polyacrylamide/calcium alginate /$EtOH_{30}$ hydrogel had a tensile strength of up to 581 kPa, that was 2.2 times greater than that of the polyacrylamide/SA hydrogel. Compressive stress-strain graphs at a strain of 10% were shown in Fig. 2.8B(c), further confirming the polyacrylamide/calcium alginate/$EtOH_{30}$ hydrogel's exceptional tensile performance. Polyacrylamide/calcium alginate /$EtOH_{30}$ hydrogels had a considerably greater elastic modulus and fracture energy compared to polyacrylamide/SA and polyacrylamide/calcium alginate hydrogels (Fig. 2.8B(d)). Notably, PAM/CA hydrogels performed mechanically just a little bit worse than PAM/SA hydrogels. This decrease may be brought on by rising radioactivity and water content. Next, ten cycles of loading-unloading at a 50% tensile strain were used to assess the fatigue resistance of the polyacrylamide/calcium alginate/$EtOH_{30}$ hydrogel. Figure 2.8B(e) illustrates what was the energy loss during the initial cycle process, but the other loading-unloading curves recorded were almost identical, demonstrating the polyacrylamide/calcium alginate/$EtOH_{30}$ hydrogel's strong fatigue resistance and cyclic hysteresis. Loading-unloading graphs were constructed across the level of 100–700% as shown in Fig. 2.8B(f), and they displayed residual stresses and hysteresis loops. Three hydrogels' dynamic rheological characteristics are shown in Fig. 2.8B(g–h). All hydrogels responded mostly elastically, as seen by the strain sweep, with Young's modulus G' surpassing loss modulus G″ in the linear viscoelastic zone (Fig. 2.8B(g)). The frequency range analysis revealed that G′ of hydrogels was frequency-independent and remained significantly higher than G″ over the whole frequency intervals (Fig. 2.8B(h)). To demonstrate the exceptional qualities of polyacrylamide/calcium alginate/$EtOH_{30}$ DN hydrogel, the mechanical strength and stretch at failure-produced materials were compiled that have previously published in the literature (Fig. 2.8B(j)).

Moreover, for the illustration of the mechanical strength of ferrogels, Yue et al. designed hydrogels with several uses as magnetic nanocomposites, made from worm-like micelles. To create nanocomposite gels, Fe_2O_3 nano-sized particles were specifically added to a temperature-sensitive block copolymer called poly (glycerol monomethacrylate)-b-poly (2-hydroxypropyl methacrylate) (PGMA-b-PHPMA) and graphene oxide (GO) dispersion at a reduced temperature of about 2 °C. Depending on the copolymer, the observed gel strengths (G′) changed when

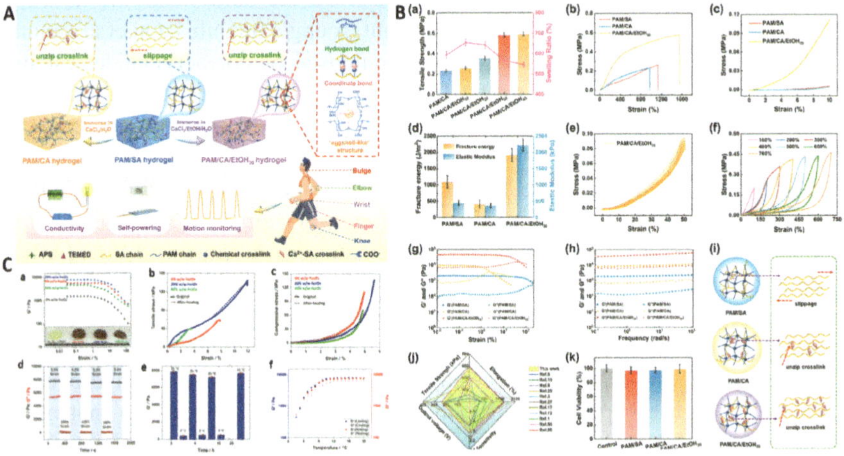

Fig. 2.8 **A** PAM/CA/$EtOH_{30}$ hydrogel design process, toughening mechanism, and applications. **B** (**a**) Radio swelling and tensile strength at varying ethanol concentrations (0–40 wt%). Stress-strain curves for tensile stress at fracture (**b**). (**d**) Fracture energy and elastic modulus of the PAM/SA, PAM/CA, and PAM/CA/$EtOH_{30}$ hydrogels. (**c**) Compressive stress-strain curves at a strain of 10%. Tensile stress-strain curves of the PAM/CA/$EtOH_{30}$ hydrogel at different strains (100–700%) and a 50% strain (**e**) are shown. G' and G'' at frequency sweep (**h**) and strain sweep (**g**). (**i**) Diagrams showing how three hydrogels dissipate energy. (**j**) Thorough analysis of the as-prepared PAM/CA/$EtOH_{30}$ material's tensile strength, break elongation, ionic conductivity, and output voltage about previously published materials. (**k**) PIECs' cell viability during a two-day incubation period in hydrogel extracts. Reproduced with permission from Ref. [99]. Copyright 2024 Elsevier. **C** (**a**) Storage modulus (G') versus percentage strain; photos of the gels under study are inset. (**b**) Room temperature tensile test results. The original gel is shown in solid lines, while the fractured gels are allowed to repair on their own in dashed lines. (**c**) Compression stress-strain curves for x% Fe_2O_3 G62–H_{170} nanocomposite worm gels at room temperature (solid lines indicate original, dotted lines indicate after self-healing) ($x = 0, 6, 20, 40$). Data on oscillatory rheology for a 20% Fe_2O_3 G_{62}–H_{170} nanocomposite worm gel are shown in (**d**). (**e**) Storage modulus dependent on temperature. (**f**) Oscillatory rheology of a 20% Fe_2O_3 G_{62}–H_{170} nanocomposite worm gel that is temperature-dependent. Reproduced with permission from Ref. [100]. Copyright 2024 American Chemical Society (Open Access under the license CC-BY-NC-ND)

the iron oxide loading was raised from 0 to 40% weight/weight. Derived from a copolymer, the measured gel strength for x% Fe_2O_3 G_{62}–H_{170} gels rose from 1.5 to 7.9 kPa as the iron oxide concentration rose from 0 to 20% w/w. However, when the Fe_2O_3 content was increased to 40% weight/weight, the observed gel strength decreased to 4.8 kPa (Fig. 2.8C(a)). Figure 2.8C(b) displays the results of tensile tests for x% Fe_2O_3 G62–H170 nanocomposite worm gels both before and following healing. The stress-strain curves of the original composite worm gels and the self-healing x% iron oxide G_{62}–H_{170} composite worm gels nearly overlapped, showing each case's fast self-healing behavior (Fig. 2.8C(c)). With a compressive modulus of around 4.5 kPa, the 20% Fe_2O_3 G_{62}–H_{170} nanocomposite worm gel, for example, had a stress of approximately 135 kPa at approximately 6% strain before breaking. Stress and strain upon self-healing were comparable to those of the initial composite worm gel (about 140 kPa, 6.2% strain, and a compressive modulus of 4.7 kPa). By varying the shear strain between 0.2 and 100% at 5-min intervals, oscillatory rheology experiments were used to examine the shear-thinning and recovery characteristics of these nanocomposite gels. With a healing effectiveness of almost 95%, the 20% w/w Fe_2O_3 G_{62}–H_{170} gel had the greatest gel strength (Fig. 2.8C(d)). Figure 2.8C(e) depicts the reversible degelation behavior across seven heating–cooling cycles, with two hours for each step. The x% Fe_2O_3 G_{62}–H_{170} nanocomposite worm gels demonstrated extremely effective reversible gelation behavior, which is comparable to the shear-thinning recovery experiments in Fig. 2.8C(d). Variable-temperature rheology investigations were used to examine the behavior of temperature response (Fig. 2.8C(f)). For x% Fe_2O_3 G_{62}–H_{170} nanocomposite worm gels, the critical gelation temperature (CGT) ranged from 2 to 9 °C during chilling and from 3 to 8 °C during heating.

2.4.3 Biocompatibility of HG-FGs

For a biological material to be used in the biomedical sector, its toxicity to cells is a critical factor. HG-FGs ought to be nontoxic and biocompatible. Essentially, biocompatibility refers to a material's capacity to interact effectively with host tissues and respond suitably in a specific environment. Two key elements of biocompatibility are biosafety and bio-functionality. The HG-FGs may become toxic if they fail to meet these criteria. HG-FG compositions made with toxic compounds frequently have issues with in vivo biocompatibility. Nonetheless, polysaccharides are accepted for use in food as they are considered non-toxic, physiologically acceptable polymers. Arshad et al. [101] created and assessed a pH-responsive nexus for the regulated administration of acyclovir using two distinct hydrophilic polymers: Quince seed gum and β-cyclodextrin (β-CD). The goal of the acute toxicity experiments was to learn more about the formed nexus's biological action. These investigations were conducted on rabbits (an animal model), which were closely observed for physical symptoms of illness, such as ocular or skin irritation, as well as other factors, including the animals' weight and food consumption. The animals exhibited no anomalies, and throughout and after the 14 days, neither the treatment nor the control groups' other metrics showed any appreciable alterations. When the rabbits were sacrificed on the fourteenth day of the trial, their vital organs were removed, and histological investigations were performed to see whether the carrier system had any harmful consequences. These organs' tissue slides were made, and they underwent microscopic inspection. As seen in Fig. 2.9a, neither group displayed significant organ abnormalities or symptoms of deterioration. Both groups' brain tissue was discovered to be normal, with oligodendrocytes, axonal cells, and well-defined neuronal cell bodies. No signs of immunological infiltration or inflammation were seen. Histopathological testing showed that the treated groups illustrated mostly normal tissue structure for major organs with no significant sign of inflammation, infection, degeneration or tissue damage. The liver injury-related hyperactivity in the portal triad region was seen in the control group. A patch of necrotic region was seen along with an accumulation of extracellular matrix. The QGP-β-CD-co-poly (methacrylate) network-treated group did not exhibit any significant pathological alterations or degeneration in key organs, demonstrating the safety and biocompatibility of the system that was designed.

As the primary second leading cause of mortality worldwide, cancer has always been a major issue. An alternate strategy for treating cancer is to use a drug delivery system to enhance the efficacy of chemical approaches. In the study conducted by Mohammadi et al. [102], the delivery of quercetin (QC), an anti-cancer medication, was examined by its incorporation into a pH-sensitive hydrogel nanocomposite consisting of $Fe_3O_4@SiO_2$ particles, agarose, and polyacrylic acid (PAA). With the use of quercetin and nanocarriers, the cytotoxicity and viability of the U-87 MG cancer cells were assessed using the MTT test. PAA, PAA-Agarose, PAA-Agarose/$Fe_3O_4@SiO_2$, PAA-Agarose/$Fe_3O_4@SiO_2$/QC, and QC were applied to the glioma

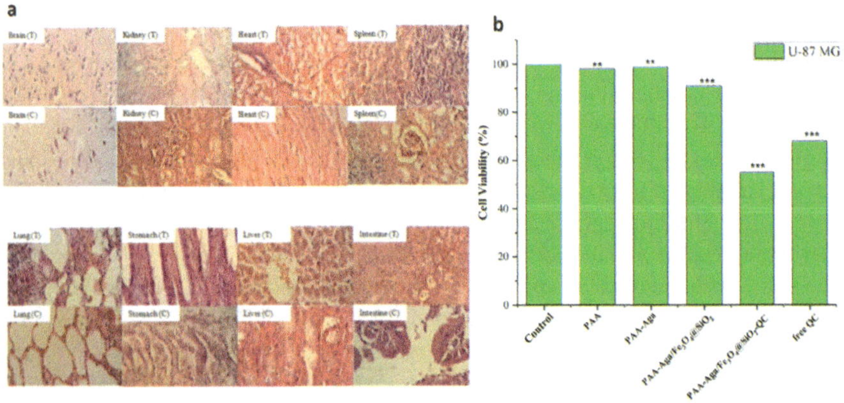

Fig. 2.9 **a** Vital organs of rabbits on histopathological slides (treated and control groups). Reproduced with permission from Ref. [101]. Copyright 2024 Springer. **b** MTT test findings during a 24-h incubation period in 96-well plates with QC-free U-87 MG treatment, PAA, PAA-Agarose, PAA-Agarose/$Fe_3O_4@SiO_2$, and PAA-Agarose/$Fe_3O_4@SiO_2$-QC treatments. Reproduced with permission from Ref. [102]. Copyright 2024 Elsevier

cell line for twenty-four hours. A set of untreated cells with 100% vitality was used as the control group. In Fig. 2.9b, the samples' cytotoxicity is contrasted with the control groups. Different levels of toxicity against U-87 MG cells are shown by the other substances, except for the control group, which displays 100% cell viability. The percentage of survival drops to 91% when $Fe_3O_4@SiO_2$ is added to the PAA-Agarose combination. There is a little increase in toxicity due to the generation of ROS and the release of Fe^{2+} ions. When SiO_2 shell is added to Fe_3O_4 nanoparticles, the proportion of cell death increases when Fe_3O_4 nanoparticles alone are used.

2.4.4 *Biodegradability*

In the biomedical sector, biodegradable HG-FGs should be utilized. "Biodegradability" denotes a living organism's capacity to decompose hydrogels into harmless byproducts. The biodegradability of HG-FGs depends on the components incorporated into the systems and the methods used for their preparation. Hydrolysis and the solubilization of biological entities from HG-FGs into final products are examples of degradation mechanisms. The HG-FGs can decompose and exit the body through bio-absorption and bio-erosion. Biodegradable polymers encompass a diverse array of both hydrophilic synthetic and natural polymers. As a result of diffusion, these polymers absorb water, swell considerably, and ultimately break down when they take in more water. Such polymers degrade according to several parameters, including molecular mass, hydrophilicity, and polymer–water interactivity. Via straightforward solubilization, other ecological variables like pH and temperature may also regulate the deterioration. This polymer class is quite restricted, though. Certain polymers can only be broken down chemically rather than simply by hydrolysis. Instead of generating hydrogels, such polymers mix with HG-FGs to generate hydrophilic monomers or polymers that may then be used to create biodegradable HG-FGs. Ester connections in these HG-FG systems are often hydrolyzed chemically, causing them to break down. Furthermore, a type of HG-FG known as polymers, such as proteins, synthetic polypeptides, and polysaccharides, may also be disintegrated by enzyme hydrolysis. These polymers have the major benefit of enzymatic breakdown, which make them suitable for application in drug delivery systems. Moreover, only glycosidase is involved in the biodegradation of polysaccharide hydrogels. Glucosamine-glucosamine, glucosamine-N-acetyl-glucosamine, and N-acetyl-glucosamine-N-acetyl-glucosamine may all be hydrolyzed by enzymes to break down chitosan. Nevertheless, redox and free radical processes also play a role in the breakdown of chitosan. Lysozyme has been employed to investigate chitosan breakdown in vitro. It was discovered that after 4 h of incubation in a pH 5.5 buffer solution, 50% acetylated chitosan had a 66% viscosity reduction [103]. The rate of degradation is not significantly altered by chitosan modification. Trimethyl chitosan degraded similarly to chitosan, although the degaradation of Trimethyl chitosan is non-sensitive with change in pH [104].

Zhang et al. [105] designed an antioxidant bioadhesive that may be applied topically and an internally implanted biodegradable hydrogel to prevent both radioactive skin damage and postoperative tumor recurrence following adjuvant radiation therapy. The biodegradable silk fibroin/perfluorocarbon (SF/PF) hydrogel loading doxorubicin (DOX) showed continuous oxygen release in the physiological environment to improve hypoxia and radiotherapy sensitivity, and simultaneous release of DOX to ultimately achieve an effective anti-cancer effect. An enzymatic degradation technique was used to ascertain the breakdown behaviors of SF/PF hydrogels at different concentrations of silk fibroin. PBS (pH 7.4, 0.1 M) was used to dissolve collagenase IV at a rate of 1.0 U/mL. After being divided into squares, the SF/PF hydrogel was weighed. After that, the samples were slowly shaken while being incubated at 37 °C in 5 mL of an enzyme solution for degradation tracing. The proportion of weight maintained about the starting weight was used to indicate quantitative changes. The weight loss of the SF/PF hydrogel reduced as the SF contents increased. Denser β-sheets probably formed because of the greater SF content concentration, which strengthened the crosslink density. By day 14, it was clear how different the SF/PF hydrogels were in terms of weight loss. After a month, the SF/PF hydrogels with SF concentrations of 3% and 4% lost weight by 68% and 47%, respectively, whereas the hydrogels with SF concentrations of 6% lost weight by 20%. Thus, the concentration of SF might be used to control the pace of deterioration of SF/PF hydrogels.

Long et al. [106] engineered the pH/magnetic dual-responsive ferrogel (Fe_3O_4@XH-Gel) by adding Xylan hemicellulose (XH), acrylic acid, poly (ethylene glycol) diacrylate, and Fe_3O_4 through graft polymerization on XH with in situ doping Fe_3O_4 MNPs started by the tetramethyl ethylenediamine/ammonium persulfate redox system. While studying the Fe_3O_4@XH-Gel ferrogels' water absorption process, equilibrium swellings were regularly attained in less than 20 h. The Fe_3O_4@XH-Gel ferrogel samples' swelling and degradation rates were compared, and the results showed that bulk erosion is the main mechanism of ferrogel degradation. The XH-Gel without Fe_3O_4 MNPs showed the highest rate of disintegration throughout the procedure, reaching virtually total breakdown after 216 h (9 days). On the other hand, the degradation rates of the Fe_3O_4@XH-Gel ferrogels containing 5 wt% and 10 wt% Fe_3O_4 MNPs were slower during the first 72 h and thereafter accelerated. This phenomenon can be explained by the ferrogel's increased density as a result of the Fe_3O_4 MNPs' inclusion. It is possible to deduce that the ferrogel is nearly entirely biodegradable in the simulated intestinal environment after taking the mass of the added Fe_3O_4 MNPs into consideration.

2.5 Summary with Scalability and Translational Aspects of HG-FGs

The basic ideas of the crosslinking techniques, polymerization approaches, bonding kinds, and, other crucial properties, which have a broad spectrum of utilization of hydrogels with ferrogels are compiled in this chapter. The significant characterization and uses of HG-FGs rely heavily on the bonding, synthesis method, and choice of polymers and crosslinking agents. The most promising solution for powering nanotheranostics is to use smart, lightweight, and flexible components. But the flexible gadgets must be strong and mechanically sustainable. Consequently, because HG-FGs can withstand significant mechanical strain and retain higher water levels in 3D frameworks that offer incredibly large surface areas for bioactive interactions, they are perfect frameworks for designing the necessary biomedical and sensing devices. HG-FGs are intrinsically flexible and ionically conductive. The other important necessity of the current era is HG-FG's ability to biodegrade. This chapter provided an overview of the development of biodegradable hydrogels and ferrogels.

High scalability is necessary to transfer HG-FGs from labs to clinics after their synthesis from monomers. The main obstacle to the development and commercialization of these materials for real-world applications is their large-scale productivity with reduced batch-to-batch variance of gels. For example, the use of iron oxide nanoparticles in medicine as contrast development in magnetic resonance imaging (MRI) has been abandoned because of poor repeatability caused by batch-to-batch size fluctuation. The healthcare industry will only adopt nanomanufacturing processes that are expedient, scalable, reproducible, and cost-effective. The large-scale synthesis of homogeneous nanomaterials is a major prerequisite for the commercialization of these nanoparticles. From in vitro cell investigations to small animals, primates, and ultimately clinical trial testing, the amount of nanomaterial required grows exponentially at each level. Therefore, the suitability of nanomaterials made utilizing scalable techniques with great control over their shape and physicochemical characteristics is essential to their effective deployment in such bio applications. The application of these nanoparticles in controlled drug delivery systems is encouraged by their large-scale synthesis and improved physicochemical and morphological characteristics. Furthermore, injectable, implantable, and sprayable HG-FGs, which have expanded over the past 60 years with significant therapeutic and social implications, are mostly covered by the translational features of HG-FGs. However, a thorough understanding of fundamental properties of HG-FGs is necessary for the advanced in vivo applications. Thus, while HG-FGs manufacture is necessary, it's also critical to consider the scalability and translational elements of hydrogels and ferrogels to facilitate their transfer from labs to clinics and enable real-world use.

References

1. X. Gan, X. Wang, Y. Huang, G. Li, H. Kang, Applications of hydrogels in osteoarthritis treatment. Biomedicine **12**, 923 (2024). https://doi.org/10.3390/biomedicines12040923
2. R. Parhi, Cross-linked hydrogel for pharmaceutical applications: a review. Adv Pharm Bull **7**, 515–530 (2017). https://doi.org/10.15171/apb.2017.064
3. J. Berger, M. Reist, J.M. Mayer, O. Felt, N.A. Peppas, R. Gurny, Structure and interactions in covalently and ionically crosslinked chitosan hydrogels for biomedical applications. Eur. J. Pharm. Biopharm. **57**, 19–34 (2004). https://doi.org/10.1016/S0939-6411(03)00161-9
4. W. SHIM, J. KIM, K. KIM, Y. KIM, R. PARK, I. KIM, I. KWON, D. LEE, pH- and temperature-sensitive, injectable, biodegradable block copolymer hydrogels as carriers for paclitaxel. Int. J. Pharm. **331**, 11–18 (2007). https://doi.org/10.1016/j.ijpharm.2006.09.027
5. J. Jin, M. Song, D.J. Hourston, Novel chitosan-based films cross-linked by Genipin with improved physical properties. Biomacromolecules **5**, 162–168 (2004). https://doi.org/10.1021/bm034286m
6. R.V. Gadhave, P.A. Mahanwar, P.T. Gadekar, Effect of glutaraldehyde on thermal and mechanical properties of starch and polyvinyl alcohol blends. Des. Monomers Polym. **22**, 164–170 (2019). https://doi.org/10.1080/15685551.2019.1678222
7. J. Cao, Y. Wu, L. Li, M. Hu, Y. Cheng, H. Zhang, J. Guo, Design of hybrid multi-crosslinked hydrogel with a combination of high stiffness, elastic applied in high temperature and high salinity reservoirs. J. Mol. Liq. **412**, 125839 (2024). https://doi.org/10.1016/j.molliq.2024.125839
8. N.-C. Zheng, H.-W. Chien, UV-crosslinking of chitosan/spent coffee ground composites for enhanced durability and multifunctionality. Int. J. Biol. Macromol. **255**, 128215 (2024). https://doi.org/10.1016/j.ijbiomac.2023.128215
9. J. Liu, K. Zhu, T. Jiao, R. Xing, W. Hong, L. Zhang, Q. Zhang, Q. Peng, Preparation of graphene oxide-polymer composite hydrogels via thiol-ene photopolymerization as efficient dye adsorbents for wastewater treatment. Colloids Surf. A Physicochem. Eng. Asp. **529**, 668–676 (2017). https://doi.org/10.1016/j.colsurfa.2017.06.050
10. A. Degirmenci, R. Sanyal, A. Sanyal, Metal-free click-chemistry: a powerful tool for fabricating hydrogels for biomedical applications. Bioconjug. Chem. **35**, 433–452 (2024). https://doi.org/10.1021/acs.bioconjchem.4c00003
11. Z.Q. Liu, Z. Wei, X.L. Zhu, G.Y. Huang, F. Xu, J.H. Yang, Y. Osada, M. Zrínyi, J.H. Li, Y.M. Chen, Dextran-based hydrogel formed by thiol-Michael addition reaction for 3D cell encapsulation. Colloids Surf. B Biointerfaces **128**, 140–148 (2015). https://doi.org/10.1016/j.colsurfb.2015.02.005
12. A. Laezza, A. Pepe, N. Solimando, F. Armiento, F. Oszust, L. Duca, B. Bochicchio, A study on Thiol-Michael addition to semi-synthetic elastin-Hyaluronan material for electrospun scaffolds. ChemPlusChem **89** (2024). https://doi.org/10.1002/cplu.202300662
13. K. Mitsuhashi, N.F. Inagaki, T. Ito, Moldable tissue-sealant hydrogels composed of *in situ* cross-linkable polyethylene glycol via thiol-Michael addition and Carbomers. ACS Biomater. Sci. Eng. **10**, 3343–3354 (2024). https://doi.org/10.1021/acsbiomaterials.3c01755
14. S. Hu, Y. Dai, L. Xin, X. Zheng, Z. Ye, S. Zhang, L. Ma, Minimally invasive delivery of human umbilical cord-derived mesenchymal stem cells by an injectable hydrogel via Diels–Alder click reaction for the treatment of intrauterine adhesions. Acta Biomater. **177**, 77–90 (2024). https://doi.org/10.1016/j.actbio.2024.02.001
15. L.J. Smith, S.M. Taimoory, R.Y. Tam, A.E.G. Baker, N. Binth Mohammad, J.F. Trant, M.S. Shoichet, Diels–Alder click-cross-linked hydrogels with increased reactivity enable 3D cell encapsulation. Biomacromolecules **19**, 926–935 (2018). https://doi.org/10.1021/acs.biomac.7b01715
16. T. Xie, H. Zhang, Y. Bai, K. Mei, Y. Zheng, C. Zhang, X. Cheng, Study on the effect of CO_2-responsive hydrogel prepared by free radical polymerization on the self-healing of cement paste fractures: application in CCUS cementing. Constr. Build. Mater. **435**, 136651 (2024). https://doi.org/10.1016/j.conbuildmat.2024.136651

17. Y. Wang, L. Li, Y. Kotsuchibashi, S. Vshyvenko, Y. Liu, D. Hall, H. Zeng, R. Narain, Self-healing and injectable shear thinning hydrogels based on dynamic Oxaborole-diol covalent cross-linking. ACS Biomater. Sci. Eng. **2**, 2315–2323 (2016). https://doi.org/10.1021/acsbiomaterials.6b00527
18. H. Zhang, F. Zhang, J. Wu, Physically crosslinked hydrogels from polysaccharides prepared by freeze–thaw technique. React. Funct. Polym. **73**, 923–928 (2013). https://doi.org/10.1016/j.reactfunctpolym.2012.12.014
19. M.M. Elsayed, Hydrogel preparation technologies: Relevance kinetics, thermodynamics and scaling up aspects. J. Polym. Environ. **27**, 871–891 (2019). https://doi.org/10.1007/s10924-019-01376-4
20. R. Du, X. Li, L.A. Fielding, Investigating the formation of polymer–nanoparticle complex Coacervate hydrogels using polymerization-induced self-assembly-derived Nanogels with a succinate-functional Core. Langmuir **40**, 20648–20656 (2024). https://doi.org/10.1021/acs.langmuir.4c02626
21. C. Wu, X. Li, X. Huang, P. Liu, J. Li, J. Liu, M. Jellico, Y. Yuan, The formation mechanism and textural properties of a complex gel based on soybean glycinin-chitosan complex coacervates: effects of pH, heat treatment temperature and centrifugation. Int. J. Biol. Macromol. **262**, 130170 (2024). https://doi.org/10.1016/j.ijbiomac.2024.130170
22. S. Awasthi, S.K. Pandey, Recent advances in smart hydrogels and carbonaceous nanoallotropes composites. Appl. Mater. Today **36**, 102058 (2024). https://doi.org/10.1016/j.apmt.2024.102058
23. S. Awasthi, J.K. Gaur, S.K. Pandey, M.S. Bobji, C. Srivastava, High-strength, strongly bonded nanocomposite hydrogels for cartilage repair. ACS Appl. Mater. Interfaces (2021). https://doi.org/10.1021/acsami.1c05394
24. S. Hu, Y. Huang, X. Liu, C. Zong, L. Lei, H. Li, Mechanically robust and highly conductive bacterial cellulose hydrogels through synergy of directional freeze–thawing and salting-out for wearable sensors. Chem. Eng. J. **499**, 156161 (2024). https://doi.org/10.1016/j.cej.2024.156161
25. I.M. El-Sherbiny, M.H. Yacoub, Hydrogel scaffolds for tissue engineering: progress and challenges. Glob. Cardiol. Sci. Pract. **2013**, 38 (2013). https://doi.org/10.5339/gcsp.2013.38
26. D. Odelli, S.R. Rodrigues, L. Silva de Sousa, N.F. Nogueira Silva, A.L. López Martínez, L.S. Queiroz, F. Casanova, A.F. de Carvalho, Effect of konjac glucomannan on heat-induced pea protein isolate hydrogels: Evaluation of structure and formation mechanisms. Food Hydrocoll. **156**, 110310 (2024). https://doi.org/10.1016/j.foodhyd.2024.110310
27. R. Gurny, E. Doelker, N.A. Peppas, Modelling of sustained release of water-soluble drugs from porous, hydrophobic polymers. Biomaterials **3**, 27–32 (1982). https://doi.org/10.1016/0142-9612(82)90057-6
28. K.-H. Wang, C.-H. Liu, D.-H. Tan, M.-P. Nieh, W.-F. Su, Block sequence effects on the self-assembly Behaviors of polypeptide-based Penta-block copolymer hydrogels. ACS Appl. Mater. Interfaces **16**, 6674–6686 (2024). https://doi.org/10.1021/acsami.3c18954
29. X. Li, J. Shi, C. Kong, C. Li, H. Wang, S. Jiang, Y. Li, Photoinitiator-grafted polymer coating enable various hydrogels with strong resistance to both drying and swelling. Chem. Eng. J. **480**, 147937 (2024). https://doi.org/10.1016/j.cej.2023.147937
30. D. Hao, Z. Wang, M. Liu, X. Guo, S. Wang, L. Jiang, Strong anchoring of hydrogels through Superwetting-assisted high-density interfacial grafting. Angew. Chem. **135** (2023). https://doi.org/10.1002/ange.202215034
31. A. Ortega, A. Sánchez, G. Burillo, Binary graft of poly (N-vinylcaprolactam) and poly(acrylic acid) onto chitosan hydrogels using ionizing radiation for the retention and controlled release of therapeutic compounds. Polymers (Basel) **13**, 2641 (2021). https://doi.org/10.3390/polym13162641
32. R. JIN, C. HIEMSTRA, Z. ZHONG, J. FEIJEN, Enzyme-mediated fast in situ formation of hydrogels from dextran–tyramine conjugates. Biomaterials **28**, 2791–2800 (2007). https://doi.org/10.1016/j.biomaterials.2007.02.032

33. A.K. Kolour, M. Shahrousvand, J. Mohammadi-Rovshandeh, D. Puppi, D. Farzaneh, Absorbable and biodegradable enzyme-crosslinked gelatin/alginate semi-IPN hydrogel wound dressings containing curcumin. Int. J. Biol. Macromol. **279**, 134938 (2024). https://doi.org/10.1016/j.ijbiomac.2024.134938
34. D. Sun, H. Wang, J. Liu, X. Wang, H. Guo, L. Xue, L. Li, J. Li, B. Zhang, Y. Xue, S. Li, Y. Liu, An enzyme cross-linked hydrogel as a minimally invasive arterial tissue sealing and anti-adhesion barrier. Nano Today **44**, 101467 (2022). https://doi.org/10.1016/j.nantod.2022.101467
35. A.B. Lugao, S.M. Malmonge, Use of radiation in the production of hydrogels. Nucl. Instrum. Methods Phys. Res. B **185**, 37–42 (2001). https://doi.org/10.1016/S0168-583X(01)00807-2
36. W.H. Wan Ishak, N.A. Rosli, I. Ahmad, S. Ramli, M.C.I. Mohd Amin, Drug delivery and in vitro biocompatibility studies of gelatin-nanocellulose smart hydrogels cross-linked with gamma radiation. J. Mater. Res. Technol. **15**, 7145–7157 (2021). https://doi.org/10.1016/j.jmrt.2021.11.095
37. R. Naranjo-Alcazar, S. Bendix, T. Groth, G. Gallego Ferrer, Research progress in enzymatically cross-linked hydrogels as injectable systems for bioprinting and tissue engineering. Gels **9**, 230 (2023). https://doi.org/10.3390/gels9030230
38. S. Awasthi, J.K. Gaur, M.S. Bobji, C. Srivastava, Nanoparticle-reinforced polyacrylamide hydrogel composites for clinical applications: a review. J. Mater. Sci. **57**, 8041–8063 (2022)
39. S. Awasthi, A. Srivastava, A. Goyal, Design strategies of magneto-responsive polyacrylamide Ferrogels. J. Clust. Sci. **36**, 214 (2025). https://doi.org/10.1007/s10876-025-02945-5
40. Y. Li, G. Huang, X. Zhang, B. Li, Y. Chen, T. Lu, T.J. Lu, F. Xu, Magnetic hydrogels and their potential biomedical applications. Adv. Funct. Mater. **23**, 660–672 (2013)
41. V.R. Shaumbwa, D. Liu, B. Archer, J. Li, F. Su, Preparation and application of magnetic chitosan in environmental remediation and other fields: a review. J. Appl. Polym. Sci. **138** (2021). https://doi.org/10.1002/app.51241
42. R.P. Dhavale, R.P. Dhavale, S.C. Sahoo, P. Kollu, S.U. Jadhav, P.S. Patil, T.D. Dongale, A.D. Chougale, P.B. Patil, Chitosan coated magnetic nanoparticles as carriers of anticancer drug Telmisartan: pH-responsive controlled drug release and cytotoxicity studies. J. Phys. Chem. Solids **148**, 109749 (2021). https://doi.org/10.1016/j.jpcs.2020.109749
43. D.T. Nguyen, N.-M. Nguyen, D.-M. Vu, M.-D. Tran, V.-T. Ta, On-demand release of drug from magnetic nanoparticle-loaded alginate beads. J. Anal. Methods Chem. **2021**, 1–7 (2021). https://doi.org/10.1155/2021/5576283
44. G.A. Kloster, D. Muraca, O. Moscoso Londoño, K.R. Pirota, M.A. Mosiewicki, N.E. Marcovich, Alginate based nanocomposites with magnetic properties. Compos. Part A Appl. Sci. Manuf. **135**, 105936 (2020). https://doi.org/10.1016/j.compositesa.2020.105936
45. N.M. Renani, N. Etesami, T. Behzad, Synthesis and characterization of novel magnetic Nano-biocomposite hydrogels based on starch-g-poly (acrylic acid) reinforced by cellulose nanofibers for Cu^{2+} ion removal. ACS Omega **8**, 21929–21940 (2023). https://doi.org/10.1021/acsomega.3c01655
46. S. Awasthi, J.K. Gaur, M.S. Bobji, Advanced Ferrogels with high magnetic response and wear resistance using carbon nanotubes. J. Alloys Compd. **848**, 156259–156269 (2020)
47. S. Fiejdasz, A. Gilarska, W. Horak, A. Radziszewska, T. Strączek, M. Szuwarzyński, M. Nowakowska, C. Kapusta, Structurally stable hybrid magnetic materials based on natural polymers—preparation and characterization. J. Mater. Res. Technol. **15**, 3149–3160 (2021). https://doi.org/10.1016/j.jmrt.2021.09.124
48. H. Ahmad, M.S. Sultana, M.A. Alam, M.M. Rahman, K. Tauer, M.A. Gafur, M.K. Sharafat, Evaluating a simple blending approach to prepare magnetic and stimuli-responsive composite hydrogel particles for application in biomedical field. Express Polym. Lett. **10**, 664–678 (2016). https://doi.org/10.3144/expresspolymlett.2016.60
49. S. Ikhaddalene, F. Zibouche, A. Ponton, A. Irekti, F. Carn, Synthesis and rheological properties of magnetic chitosan hydrogel. Period. Polytech. Chem. Eng. **65**, 378–388 (2021). https://doi.org/10.3311/PPch.17148

50. S. Zhang, H. Gao, P. Guo, T. Li, T. Lin, R. Ding, Z. Wang, P. He, Fabrication and extrusion of the PAAm-SAlg hydrogels with magnetic particles. Colloids Surf. A Physicochem. Eng. Asp. **603**, 125280 (2020). https://doi.org/10.1016/j.colsurfa.2020.125280
51. G.R. Mahdavinia, M. Soleymani, H. Etemadi, M. Sabzi, Z. Atlasi, Model protein BSA adsorption onto novel magnetic chitosan/PVA/laponite RD hydrogel nanocomposite beads. Int. J. Biol. Macromol. **107**, 719–729 (2018). https://doi.org/10.1016/j.ijbiomac.2017.09.042
52. S. Farzaneh, S. Hosseinzadeh, R. Samanipour, S. Hatamie, J. Ranjbari, A. Khojasteh, Fabrication and characterization of cobalt ferrite magnetic hydrogel combined with static magnetic field as a potential bio-composite for bone tissue engineering. J. Drug Deliv. Sci. Technol. **64**, 102525 (2021). https://doi.org/10.1016/j.jddst.2021.102525
53. R. Messing, N. Frickel, L. Belkoura, R. Strey, H. Rahn, S. Odenbach, A.M. Schmidt, Cobalt ferrite nanoparticles as multifunctional cross-linkers in PAAm Ferrohydrogels. Macromolecules **44**, 2990–2999 (2011). https://doi.org/10.1021/ma102708b
54. R. Barbucci, D. Pasqui, G. Giani, M. De Cagna, M. Fini, R. Giardino, A. Atrei, A novel strategy for engineering hydrogels with ferromagnetic nanoparticles as crosslinkers of the polymer chains. Potential applications as a targeted drug delivery system. Soft Matter **7**, 5558 (2011). https://doi.org/10.1039/c1sm05174a
55. W. Haas, M. Zrinyi, H.-G. Kilian, B. Heise, Structural analysis of anisometric colloidal iron(III)-hydroxide particles and particle-aggregates incorporated in poly(vinyl-acetate) networks. Colloid Polym. Sci. **271**, 1024–1034 (1993). https://doi.org/10.1007/BF00659291
56. S.G. Starodubtsev, E.V. Saenko, M.E. Dokukin, V.L. Aksenov, V.V. Klechkovskaya, I.S. Zanaveskina, A.R. Khokhlov, Formation of magnetite nanoparticles in poly(acrylamide) gels. J. Phys. Condens. Matter **17**, 1471–1480 (2005). https://doi.org/10.1088/0953-8984/17/10/004
57. N.R. Nagireddy, M.M. Yallapu, V. Kokkarachedu, R. Sakey, V. Kanikireddy, J. Pattayil Alias, M.R. Konduru, Preparation and characterization of magnetic nanoparticles embedded in hydrogels for protein purification and metal extraction. J. Polym. Res. **18**, 2285–2294 (2011). https://doi.org/10.1007/s10965-011-9642-2
58. L. Mei, J. Cong, S. Li, T. Ren, Y. Bai, Q. Liu, C. Miao, W. Ding, T. Luo, Polyacrylamide-chitosan based magnetic hydrogels with high stiffness and ultra-toughness. Compos. Part A Appl. Sci. Manuf. **168**, 107478 (2023). https://doi.org/10.1016/j.compositesa.2023.107478
59. M. Simeonov, A.A. Apostolov, M. Georgieva, D. Tzankov, E. Vassileva, Poly(acrylic acid-co-acrylamide)/polyacrylamide pIPNs/magnetite composite hydrogels: synthesis and characterization. Gels **9**, 365 (2023). https://doi.org/10.3390/gels9050365
60. Y. Wang, B. Li, Y. Zhou, D. Jia, In situ mineralization of magnetite nanoparticles in chitosan hydrogel. Nanoscale Res. Lett. **4**, 1041 (2009). https://doi.org/10.1007/s11671-009-9355-1
61. N. Yuan, L. Xu, L. Zhang, H. Ye, J. Zhao, Z. Liu, J. Rong, Superior hybrid hydrogels of polyacrylamide enhanced by bacterial cellulose nanofiber clusters. Mater. Sci. Eng. **67**, 221–230 (2016)
62. A. Alam, H.C. Kuan, Z. Zhao, J. Xu, J. Ma, Novel polyacrylamide hydrogels by highly conductive, water-processable graphene. Compos. Part A Appl. Sci. Manuf. **93**, 1–9 (2017)
63. E. Olăreț, Ș.I. Voicu, R. Oprea, F. Miculescu, L. Butac, I.C. Stancu, A. Serafim, Nanostructured polyacrylamide hydrogels with improved mechanical properties and antimicrobial behavior. Polymers (Basel) **14** (2022)
64. Y. Ohsedo, M. Taniguchi, K. Saruhashi, H. Watanabe, Improved mechanical properties of polyacrylamide hydrogels created in the presence of low-molecular-weight hydrogelators. RSC Adv. **5**, 90010–90013 (2015)
65. G. Filipcsei, I. Csetneki, A. Szilágyi, M. Zrínyi, *Magnetic Field-Responsive Smart Polymer Composites* (2007), pp. 137–189. https://doi.org/10.1007/12_2006_104
66. M.F. Akhtar, M. Hanif, N.M. Ranjha, Methods of synthesis of hydrogels … a review. Saudi Pharm. J. **24**, 554–559 (2016). https://doi.org/10.1016/j.jsps.2015.03.022
67. N.D. Tsihlis, J. Murar, M.R. Kapadia, S.S. Ahanchi, C.S. Oustwani, J.E. Saavedra, L.K. Keefer, M.R. Kibbe, Isopropylamine NONOate (IPA/NO) moderates neointimal hyperplasia following vascular injury. J. Vasc. Surg. **51**, 1248–1259 (2010). https://doi.org/10.1016/j.jvs.2009.12.028

68. N. Annabi, A. Tamayol, J.A. Uquillas, M. Akbari, L.E. Bertassoni, C. Cha, G. Camci-Unal, M.R. Dokmeci, N.A. Peppas, A. Khademhosseini, 25th anniversary article: rational design and applications of hydrogels in regenerative medicine. Adv. Mater. **26**, 85–124 (2014). https://doi.org/10.1002/adma.201303233
69. E.M. Ahmed, Hydrogel: preparation, characterization, and applications: a review. J. Adv. Res. **6**, 105–121 (2015). https://doi.org/10.1016/j.jare.2013.07.006
70. J.M. Seidel, S.M. Malmonge, Synthesis of polyHEMA hydrogels for using as biomaterials. Bulk and solution radical-initiated polymerization techniques. Mater. Res. **3**, 79–83 (2000). https://doi.org/10.1590/S1516-14392000000300006
71. W.E. Hennink, C.F. van Nostrum, Novel crosslinking methods to design hydrogels. Adv. Drug Deliv. Rev. **54**, 13–36 (2002). https://doi.org/10.1016/S0169-409X(01)00240-X
72. T. Billiet, M. Vandenhaute, J. Schelfhout, S. Van Vlierberghe, P. Dubruel, A review of trends and limitations in hydrogel-rapid prototyping for tissue engineering. Biomaterials **33**, 6020–6041 (2012). https://doi.org/10.1016/j.biomaterials.2012.04.050
73. A.-A.A. Abdel-Azim, M.S. Farahat, A.M. Atta, A.A. Abdel-Fattah, Preparation and properties of two-component hydrogels based on 2-acrylamido-2-methylpropane sulphonic acid. Polym. Adv. Technol. **9**, 282–289 (1998). https://doi.org/10.1002/(SICI)1099-1581(199805)9:5<282::AID-PAT755>3.0.CO;2-N
74. N. Ranganathan, R. Joseph Bensingh, M. Abdul Kader, S.K. Nayak, *Synthesis and Properties of Hydrogels Prepared by Various Polymerization Reaction Systems* (2018), pp. 1–25. https://doi.org/10.1007/978-3-319-76573-0_18-1
75. C.E. Carraher Jr., *Introduction to Polymer Chemistry* (CRC Press, 2012). https://doi.org/10.1201/b13684
76. D.S. Achilias, G.D. Verros, Modeling of diffusion-controlled reactions in free radical solution and bulk polymerization: model validation by DSC experiments. J. Appl. Polym. Sci. **116**, 1842–1856 (2010). https://doi.org/10.1002/app.31675
77. P. Sikdar, M.M. Uddin, T.M. Dip, S. Islam, M.S. Hoque, A.K. Dhar, S. Wu, Recent advances in the synthesis of smart hydrogels. Mater. Adv. **2**, 4532–4573 (2021). https://doi.org/10.1039/D1MA00193K
78. J.R. Ebdon, *Introduction to Polymers*, 2nd edn, ed. by R.J. Young, P.A. Lovell (Chapman and Hall, London, 1991), pp. 443. ISBN 0-412-30640-9 (PB); ISBN 0-412-30630-1 (HB), Polym. Int. **27**, 207–208 (1992). https://doi.org/10.1002/pi.4990270217
79. M. Chanda, *Introduction to Polymer Science and Chemistry* (CRC Press, 2006). https://doi.org/10.1201/9781420007329
80. M. Liu, R. Liang, F. Zhan, Z. Liu, A. Niu, Preparation of superabsorbent slow release nitrogen fertilizer by inverse suspension polymerization. Polym. Int. **56**, 729–737 (2007). https://doi.org/10.1002/pi.2196
81. C.E. Carraher, *Carraher's Polymer Chemistry*, 10th edn. (CRC Press, 2017). https://doi.org/10.1201/9781315116600
82. R.J. Young, P.A. Lovell, *Introduction to Polymers* (CRC Press, 2011). https://doi.org/10.1201/9781439894156
83. H.A. Essawy, M.B.M. Ghazy, F.A. El-Hai, M.F. Mohamed, Superabsorbent hydrogels via graft polymerization of acrylic acid from chitosan-cellulose hybrid and their potential in controlled release of soil nutrients. Int. J. Biol. Macromol. **89**, 144–151 (2016). https://doi.org/10.1016/j.ijbiomac.2016.04.071
84. M.W. Tibbitt, A.M. Kloxin, L.A. Sawicki, K.S. Anseth, Mechanical properties and degradation of chain and step-polymerized photodegradable hydrogels. Macromolecules **46**, 2785–2792 (2013). https://doi.org/10.1021/ma302522x
85. D. Buenger, F. Topuz, J. Groll, Hydrogels in sensing applications. Prog. Polym. Sci. **37**, 1678–1719 (2012). https://doi.org/10.1016/j.progpolymsci.2012.09.001
86. W. Guo, C. Lu, R. Orbach, F. Wang, X. Qi, A. Cecconello, D. Seliktar, I. Willner, pH-stimulated DNA hydrogels exhibiting shape-memory properties. Adv. Mater. **27**, 73–78 (2015). https://doi.org/10.1002/adma.201403702

87. J. Sun, G. Jiang, Y. Wang, F. Ding, Thermosensitive chitosan hydrogel for implantable drug delivery: Blending PVA to mitigate body response and promote bioavailability. J. Appl. Polym. Sci. **125**, 2092–2101 (2012). https://doi.org/10.1002/app.36297
88. S.J. Kim, K.J. Lee, I.Y. Kim, S.I. Kim, Swelling kinetics of interpenetrating polymer hydrogels composed of poly(vinyl alcohol)/chitosan. J. Macromol. Sci. A **40**, 501–510 (2003). https://doi.org/10.1081/MA-120019888
89. C.-J. Wu, A.K. Gaharwar, P.J. Schexnailder, G. Schmidt, Development of biomedical polymer-silicate nanocomposites: a materials science perspective. Materials **3**, 2986–3005 (2010). https://doi.org/10.3390/ma3052986
90. J. Liu, C. Du, W. Huang, Y. Lei, Injectable smart stimuli-responsive hydrogels: pioneering advancements in biomedical applications. Biomater. Sci. **12**, 8–56 (2024). https://doi.org/10.1039/D3BM01352A
91. S. Awasthi, Ferrogels towards nanotheranostics. Mater. Today Chem. **35**, 101877 (2024)
92. S. Bashir, M. Hina, J. Iqbal, A.H. Rajpar, M.A. Mujtaba, N.A. Alghamdi, S. Wageh, K. Ramesh, S. Ramesh, Fundamental concepts of hydrogels: synthesis, properties, and their applications. Polymers (Basel) **12**, 2702 (2020). https://doi.org/10.3390/polym12112702
93. B.S.J. Charles, V. Kumaravel, S.K. Sadasivam, Biogenic synthesis and characterization of pH-sensitive magnetic PVA/starch/$MnFe_2O_4$ hydrogel for prolonged release of 5-fluorouracil. J. Inorg. Organomet. Polym. Mater. (2024). https://doi.org/10.1007/s10904-024-03175-4
94. S. Bashir, Y.Y. Teo, S. Ramesh, K. Ramesh, M.W. Mushtaq, Rheological behavior of biodegradable N-succinyl chitosan-g-poly (acrylic acid) hydrogels and their applications as drug carrier and in vitro theophylline release. Int. J. Biol. Macromol. **117**, 454–466 (2018). https://doi.org/10.1016/j.ijbiomac.2018.05.182
95. X. Li, J.P. Gong, Design principles for strong and tough hydrogels. Nat. Rev. Mater. **9**, 380–398 (2024). https://doi.org/10.1038/s41578-024-00672-3
96. M.L. Oyen, Mechanical characterisation of hydrogel materials. Int. Mater. Rev. **59**, 44–59 (2014). https://doi.org/10.1179/1743280413Y.0000000022
97. J.W. Goodwin, R.W. Hughes, *Rheology for Chemists, Royal Society of Chemistry* (Cambridge, 2008). https://doi.org/10.1039/9781847558046
98. A. Jobling, *An introduction to Rheology*, ed. by H.A. Barnes, J.F. Hutton, K. Walters (Elsevier Science Publishers, Amsterdam, 1989). pp. v + 199, ISBN 0-444-87469-0, Polym. Int. **25**, 61–61 (1991). https://doi.org/10.1002/pi.4990250112
99. K. Ou, M. Wang, C. Meng, K. Guo, N. Shariar Emon, J. Li, K. Qi, Y. Dai, B. Wang, Enhanced mechanical strength and stretchable ionic conductive hydrogel with double-network structure for wearable strain sensing and energy harvesting. Compos. Sci. Technol. **255**, 110732 (2024). https://doi.org/10.1016/j.compscitech.2024.110732
100. Q. Yue, S. Wang, S.T. Jones, L.A. Fielding, Multifunctional self-assembled block copolymer/iron oxide nanocomposite hydrogels formed from wormlike micelles. ACS Appl. Mater. Interfaces (2024). https://doi.org/10.1021/acsami.4c03007
101. S. Arshad, A. Mahmood, U. Rehman, H. Ijaz, R.M. Sarfraz, Z. Ahmad, M. Shafeeq Ur Rahman, Development and characterization of stimuli responsive quince gum/β-CD grafted poly (methacrylate) hydrogels for controlled delivery of acyclovir. Polym. Bull. **81**, 9737–9774 (2024). https://doi.org/10.1007/s00289-024-05167-3
102. Z.S. Mohammadi, M. Pourmadadi, M. Abdouss, S.H. Jafari, A. Rahdar, A.M. Díez-Pascual, pH-sensitive polyacrylic acid/Fe3O4@SiO2 hydrogel nanocomposite modified with agarose for controlled release of quercetin. Inorg. Chem. Commun. **163**, 112338 (2024). https://doi.org/10.1016/j.inoche.2024.112338
103. H. Onishi, Y. Machida, Biodegradation and distribution of water-soluble chitosan in mice. Biomaterials **20**, 175–182 (1999). https://doi.org/10.1016/S0142-9612(98)00159-8
104. R.J. Verheul, M. Amidi, M.J. van Steenbergen, E. van Riet, W. Jiskoot, W.E. Hennink, Influence of the degree of acetylation on the enzymatic degradation and in vitro biological properties of trimethylated chitosans. Biomaterials **30**, 3129–3135 (2009). https://doi.org/10.1016/j.biomaterials.2009.03.013

105. Z. Zhang, Q. Cao, Y. Xia, C. Cui, Y. Qi, Q. Zhang, Y. Wu, J. Liu, W. Liu, Combination of biodegradable hydrogel and antioxidant bioadhesive for treatment of breast cancer recurrence and radiation skin injury. Bioact. Mater. **31**, 408–421 (2024). https://doi.org/10.1016/j.bioactmat.2023.08.021
106. J. Long, G. Zhou, X. Yu, J. Xu, L. Hu, A. Pranovich, Q. Yong, Z.-H. Xie, C. Xu, Harnessing chemical functionality of xylan hemicellulose towards carbohydrate polymer-based pH/magnetic dual-responsive nanocomposite hydrogel for drug delivery. Carbohydr. Polym. **343**, 122461 (2024). https://doi.org/10.1016/j.carbpol.2024.122461

Chapter 3
Classification of Hydrogels and Ferrogels

3.1 Classification of Hydrogels

Physical characteristics, swelling type, production technique, ionic charges, origin, sources, observed crosslinking type, and rate of biodegradation all play a role in the categorization of hydrogels. Although Fig. 3.1 makes it evident that the specifics of each type's classification are beyond the scope of this chapter, some of the well-known hydrogels that draw significant interest from scientists are included.

3.1.1 Hydrogels Based on their Source or Origin

In general, hydrogels may be classified as either natural or synthetic depending on their origin. Designing hydrogels from natural biomass resources has advanced significantly in recent years. Generally, proteins originating from biological origins (e.g., collagen, elastin, gelatin, fibrin, etc.) or polysaccharides (e.g., chitosan, hyaluronic acid, alginate, agarose, cellulose, etc.) are used to create natural hydrogels. Polysaccharides are sourced from the shells of sea crustaceans, marine algae, and plants for hydrogel manufacturing. In contrast, proteins such as elastin and collagen are naturally occurring components of the extracellular matrix (ECM) and are frequently produced from human or animal sources (such as the placenta or tendons/skin). In light of global energy problems and environmental concerns, polysaccharides have been the subject of substantial biomedical study due to their non-toxic nature, low cost, biocompatibility, and biodegradability. They also have carboxyl acid (COOH), hydroxyls (OH), or amine (NH_2) groups, which offer an important platform for hydrogen bonding, functionalization, anchoring with additional groups, or crosslinking-enabling chemical alterations.

S. Awasthi, *Smart Hydrogels and Ferrogels*, Materials Horizons: From Nature to Nanomaterials, https://doi.org/10.1007/978-981-95-9077-3_3

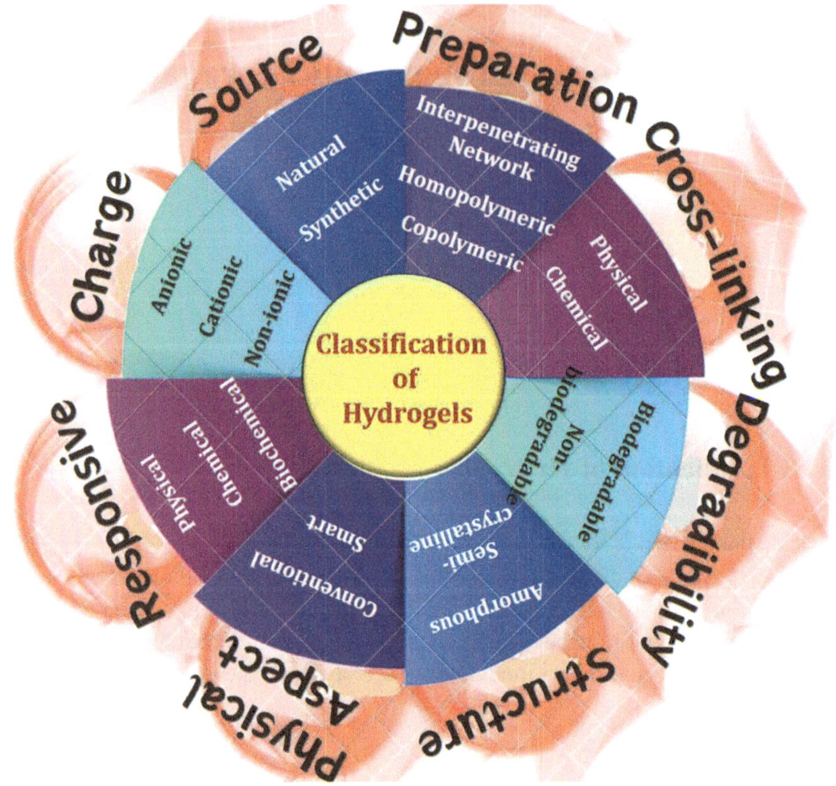

Fig. 3.1 Schematic presentation of the classification of hydrogels based on various parameters

Natural hydrogels are of low-toxicity, biodegradable with high biocompatibility. Natural hydrogels, which are produced from the extracellular matrix, possess molecular architectures that naturally promote adhesion and proliferation of cells. Other hydrogels made from plant-based components are easily accessible and do not contain any risk of animal-derived viruses. However, these materials' unclear structures make it difficult to manage their mechanical qualities (such as stiffness and flexibility), and their usage in many biomedical applications has been limited due to their challenging repeatability in large-scale manufacture [1].

Conversely, pre-structured hydrogels, which are based on the structure of monomers or polymers are referred to as synthetic hydrogels. In addition to other polymers, PEG, polyvinyl alcohol (PVA), polyethylene oxide (PEO), poly(methacrylic acid), poly(acrylamide) (PAM), and poly(N-isopropylacrylamide) (PIPAM) are frequently used. Higher water absorption capacity, improved physical and chemical stability, repeatability, and higher gel strength are some advanced properties that synthetic hydrogels may possess due to their chemically crosslinked structure. These properties are critical for developing scaffolds for bone implants, drug release systems, tissue engineering, biosensors, etc. Besides these benefits, synthetic hydrogels have various structural chemistries that might result in poor biological activity and compatibility and are not always suitable for cells or tissues. The majority of current research efforts are concentrated on creating biohybrid hydrogels, which combine the fabrication of synthetic polymers with the natural polymers, to get beyond these restrictions. Generally, several variables, including chemical stability, biocompatibility, biological activity, mechanical properties, toxicity, availability, cost, swelling ratio, biodegradability, etc., determine whether using synthetic, natural, or composite hydrogels for a perticular clinical application.

3.1.2 Hydrogel Classification Related to their Preparation Method

Depending on how they are prepared, hydrogels can be generically categorized as homopolymeric, copolymeric, or interpenetrating polymer network (IPN) hydrogels [2]. While copolymer hydrogels comprise multimonomeric polymers with at least one hydrophilic polymer, organized in block, random, or alternating patterns, homopolymeric hydrogels are polymer networks made from one polymer containing identical repeating monomers. In contrast, IPN hydrogels are made up of two naturally or artificially crosslinked polymers that are contained inside a network structure. Semi-IPN occurs when one component is crosslinked while the other is not. An injectable hydrogel was created by Pescosolido et al. [3] using the IPN of two polysaccharides, namely dextran-HEMA and calcium alginate. The IPN hydrogels showed promising properties for tissue engineering applications and targeted drug administration, and they were fully biodegradable. Additional efforts were undertaken to improve polysaccharide stability by semi-IPN hybridization with PVA to broaden the range of these polysaccharides [4]. Wang et al. [5] created a novel gelatin-alginate IPN hydrogel using physical crosslinking more recently, and it had a 79% water content. When compared to pure gelatin hydrogel, the resulting gel showed much better mechanical qualities. Awasthi et al. [6] fabricated synthetic polyacrylamide hydrogels with encapsulation of Ag and hexagonal boron nitride (hBN). The optimum concentration for Ag was observed to be 0.1 wt. % while for hBN, it was 0.3 wt. %. The compressive strength, durability, biocompatibility, and binding energy for the trimer sample (PAM-Ag-hBN) were reported to be the highest.

3.1.3 Crosslinking-Based Hydrogels

Hydrophilic polymer chains can be crosslinked in a variety of ways to create stable hydrogel polymer networks. When hydrogels are swelled, the crosslinking helps maintain their three-dimensional structure and controls water absorption [7]. The two most commonly used crosslinking techniques for creating hydrogels are physical and chemical crosslinking. Without the use of crosslinking chemicals that might be harmful to cells or tissues, physically crosslinked hydrogels can be created in extremely mild conditions. Hydrogen bonding, ionic/electrostatic interactions, charge interactions, freezing-thawing, stereo-complexing, hydrophobic interactions, protein contacts, and crystallization are some of the methods used to create physically crosslinked hydrogels [8]. Compared to physically crosslinked hydrogels, the covalently crosslinked network of chemically crosslinked hydrogels contains stronger and often more stable connections. There are several methods for creating chemical crosslinks, including the copolymerization of multifunctional monomers, crosslinker reactions, high-energy radiation applications, pendant group chemical processes, etc. [9, 10]. The most often employed techniques are the following: enzyme-enabled crosslinking reactions [11], Schiff base crosslinking reactions [12], Diels–Alder click reaction [13], oxime reaction [14], Michael addition [15], and photo-, redox-, thermal-, or radiation-initiated free-radical polymerization [16].

Considering their crosslinked chain charge, hydrogels may be separated into four categories: ampholytic, cationic, neutral, and anionic. The charge on each of the individual polymers that make up the network structure determines the charge of the entire network. The many crosslinking techniques that have been used recently to create hydrogels are depicted in Fig. 3.2.

3.1.3.1 Electrostatic or Ionic Interactions for Crosslinking

In general, the molecular bond between the cationic and anionic polyelectrolytes causes ionic or electrostatic crosslinking. To create chitosan-based hydrogels, for instance, phosphate groups (negatively charged) of glycerol phosphate disodium salt and the amine groups of chitosan (positively charged), a naturally occurring polymer, can electrostatically crosslink [18]. Similarly, divalent cations such as magnesium (Mg^{2+}), calcium (Ca^{2+}), barium (Ba^{2+}), and others were used to crosslink alginate, an anionic polymer made of glucuronic acid and mannuronic acid monomers [19]. Because crosslinking processes are simple, this technique is widely used to encapsulate proteins and medications [20]. A novel injectable DNA hydrogel created by electrostatic interaction between the amino group of chitosan (positively charged) and the DNA molecule (negatively charged) was described by Fanghao et al. [21]. These biocompatible hydrogels were easy and reasonably priced to create. Recently, Wang et al. [22] and Lei et al. [23] synthesized self-healing hydrogels with electrostatic crosslinking. In former research, a previously published procedure was used to prepare N-acryloyl-1,2-diaminoethane hydrochloride (ADE) [24]. Then, using a Michael addition between the vinyl double bond of another monomer molecule and the amino (or ammonium) groups of ADE, the hyperbranched macromonomer poly(N-acryloyl-1,2-diaminoethane hydrochloride) (HADE) was created (Fig. 3.3A a) [24, 25]. To create hydrogels cross-linked with ionic electrostatic interactions (IEC gel), HADEs and AAm were copolymerized in situ while polyacrylate sodium (PAANa) and different salt concentrations were present. IEC gels were directly

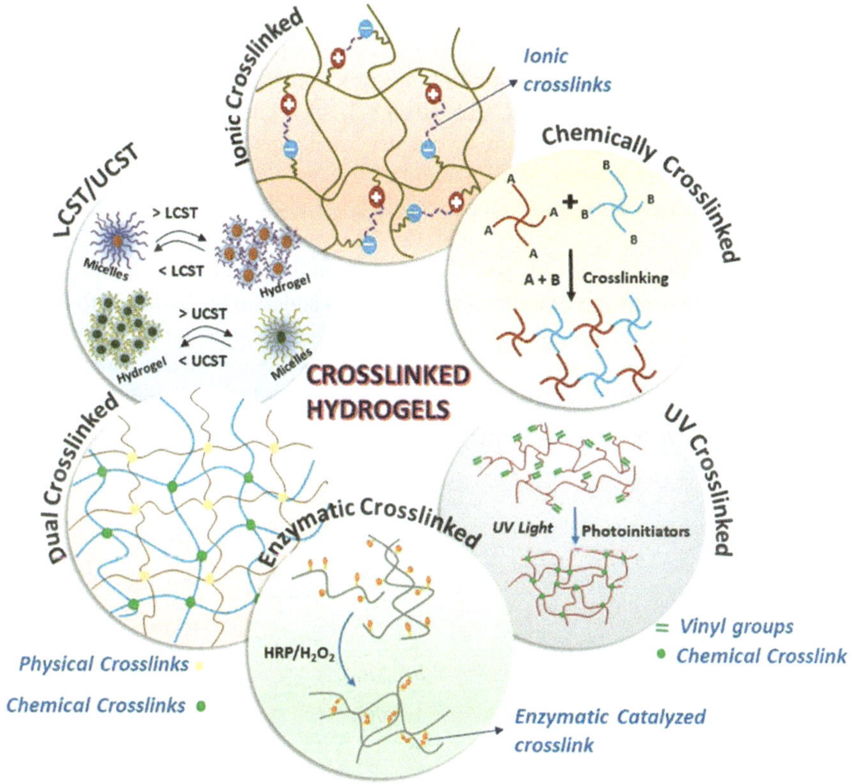

Fig. 3.2 Techniques for crosslinking were used to make the hydrogel. Physical crosslinking (ionic crosslinking), chemical crosslinking, LCST/UCST, UV-based crosslinking, dual crosslinking, and enzyme crosslinking are some of the most widely used crosslinking techniques that are depicted in the picture. Reproduced with permission from Ref. [17]. Copyright 2023 Royal Society of Chemistry (Open Access)

created during polymerization by controlling the proportion of salt and polyelectrolyte, which promoted electrostatic cross-linking between the anionic carboxylate groups and cationic hyperbranched units (Fig. 3.3A b). This technique made it possible to use several initiators during polymerization to create IEC gels with various structures. Route 1 involved the synthesis of single electrostatically cross-linked IEC gels with 2,2′-azobis[2-(2-imidazolin-2-yl)propane] dihydrochloride (VA-044) as the initiator. The creation of the resultant IEC gel network, known as S-IEC gels, was entirely dependent on the electrostatic interactions between the anionic PAANa and the cationic hyperbranched copolymers. On the other hand, tert-butyl hydroperoxide (TBHP) or potassium persulfate (KPS) formed redox pairs with the amino groups of HADE when they were employed as the initiator [26–28]. While radicals in the reaction mixture either started the homopolymerization of monomers or stopped expanding polymer chains, creating a chemically cross-linking network, the radicals on the hyperbranched copolymer chains started the grafting polymerization of monomers. Surplus amino and carboxylate groups were used to accomplish additional electrostatic physical crosslinking, producing dual (chemical and electrostatic physical) cross-linked IEC gels (Route 2), also known as D-IEC gels. IEC gels lacking hyperbranched structures were also created as a reference by replacing HADE with ADE, resulting in SL-IEC and DL-IEC gels.

In another research, authors developed an effective adsorbent using biomass materials [23]. First, a single-network hydrogel (CS-SA) was successfully created by using the sodium alginate (SA) and chitosan (CS) electrostatic interaction. Calcium phytate (PCa), a naturally occurring cross-linking agent, was added to the hydrogel to further improve its mechanical strength and chemical stability. A composite hydrogel with a twofold improved network structure (CS-SA/PCa) was created by weaving the original CS-SA network with the recently created cross-linked network while PCa was present (Fig. 3.3B). According to experimental findings, the combination of CS and

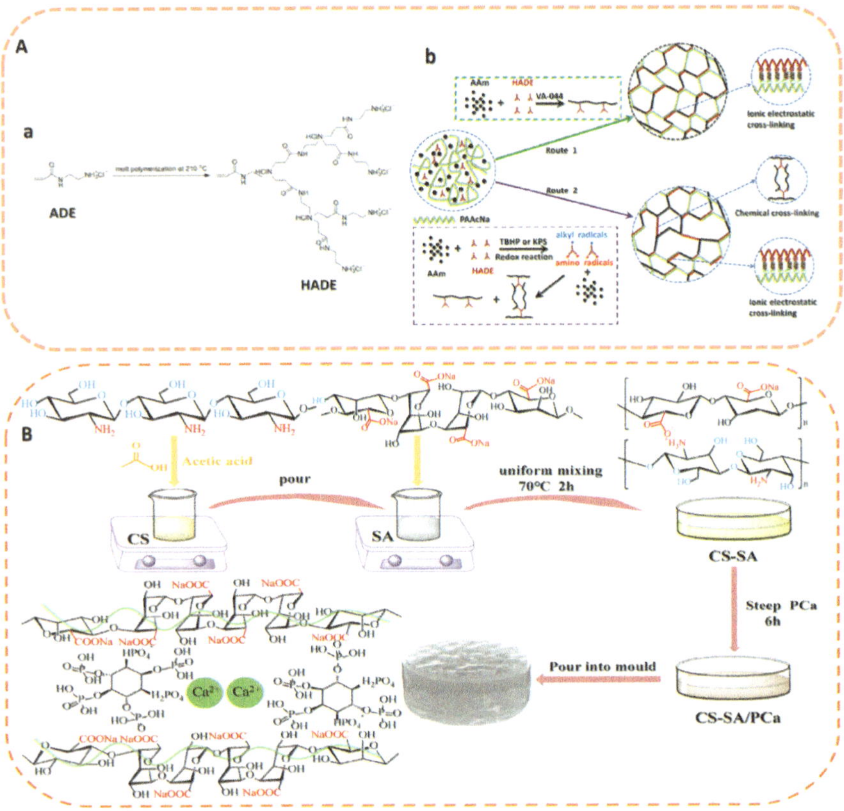

Fig. 3.3 **A** Production pathways for S-IEC and D-IEC gels. Reproduced with permission from Ref. [22]. Copyright 2025 Elsevier. **B** Schematic to show the synthesis of hydrogel. Reproduced with permission from Ref. [23]. Copyright 2025 Elsevier

SA gives the original hydrogel a high degree of structural stability, and the inclusion of PCa improves its mechanical characteristics and successfully reduces the possibility of secondary contamination from conventional harmful chemical cross-linking agents. Furthermore, the PCa molecule's phosphate group is abundant in negative charge centers that can create coordination chemical bonds with U(VI) through oxygen atoms and electrostatically interact with U(VI).

3.1.3.2 Hydrophobic Interactions to Crosslink

A key factor in creating high strength hydrogels for large biological systems is hydrophobic interactions. By adding hydrophobic structural units to the hydrophilic polymer chain, hydrophobic interactions can be created. Because of the flexible mobility of junction zones in the hydrogel network, which aids in energy dissipation and increases fracture toughness, hydrogels generated via hydrophobic contacts often display high toughness [29]. Associative thickeners such as hydrophobically ethoxylated urethanes have made extensive use of hydrophilic interactions [30]. In an aqueous solution, the hydrophobic groups in these polymers have the ability to self-organize into micelles that resemble flowers, joining hydrophilic groups. Two primary techniques are used to create hydrophobic interactions: ultrasonic treatment and lower critical solution temperature (LCST) or higher critical solution temperature (UCST). The process of repeatedly freezing and thawing PVA hydrogels is the most widely used example. The crystalline nature of the PVA structure is altered when the temperature drops below a specific critical temperature, and the PVA solution transforms into a hydrogel as a result of the intermolecular hydrogen bonding that takes place when the temperature drops below the UCST [31]. The external temperature change has a significant impact on a number of characteristics, including the gelation time, degree of swelling, and water affinity [32]. The primary functions of UCST hydrogels are hydrophilic, and at high temperatures, they expand excessively (in the right solvents). Natural polymers such as gelatin, collagen, agarose, carrageenan, etc., are examples of UCST hydrogels [33]. On the other hand, because of the harmonious balance between hydrophobic effects and hydrogen bonding, certain polysaccharides/macromers transform into hydrogels when the temperature rises over the LCST [34]. LCST hydrogels experience sol-to-gel transitions based on temperature and have both hydrophilic and hydrophobic functions in their chain. Tuncaboylu et al. [35] carried out and documented several tests about the creation of PAM hydrogels with high toughness through reversible hydrophobic interactions. Using acrylate-terminated poly(ε-caprolactone)–poly(ethylene glycol)–poly(ε-caprolactone)[Ac(PCEC)] as the hydrophobic unit, Deng et al. [36] used this approach to create stretchy, conducting hydrogels. These hydrogels showed promise for use in wearable sensors, electronic skin, and intelligent robots, among other useful devices.

Cao et al. [37] reported that hydrophobic interactions and hydrogen bond interactions were the main forces behind the development of peptide hydrogels. The researchers provided a more fundamental knowledge of this process by demonstrating the beneficial impact of amino acid molecular properties and organization on hydrogel formation using proteomics in conjunction with molecular docking simulations. In conclusion, the work improved the knowledge of the basic processes behind the synthesis of peptide hydrogels that self-assemble from plant proteins while also establishing an effective strategy for doing so.

3.1.3.3 Crosslinking Through Enzyme Catalysis

Another method that is becoming more and more popular is enzymatic crosslinking, which allows to control the properties of the enzyme and hence control the gel formation. Numerous factors, such as a particular enzyme type, their structural arrangement, physiological circumstances, etc., affect how gels develop. Enzyme-catalyzed hydrogel production is simple and done under moderate physiological circumstances as contrary to physical and chemical crosslinking techniques. For instance, the crosslinking method does not use any hazardous radiation, high temperatures, poisonous chemicals, or great efficiency. Most of the enzymes now used in the crosslinking process are comparable to those involved in catalytic processes within the body. Furthermore, substrate-specific enzyme reactions prevent other harmful effects and unanticipated byproducts that occur via chemical or photo-crosslinking techniques. Oxidation processes, which are mediated by enzymes like tyrosinase or peroxidase, change substrates into reactive forms that may eventually form covalent connections. Horseradish peroxidase (HRP) and transglutaminase (TG) are the two most often utilized enzymes [38]. Through the efficient coupling of hydroxyphenylpropionic acid moieties [39], HRP may form networks among polymers. In the presence of H_2O_2 [40], HRP catalyzes the coupling of tyramine, phenol [41], aniline [42], and Bi et al. [43] recently created injectable hydrogels for tissue engineering purposes by employing an enzymatically crosslinkable carboxymethyl chitin conjugate modified by tyramine. HRP was used to create the hydrogels under moderate physiological conditions (with H_2O_2). According to both in vitro and in vivo research, enzymatically crosslinked hydrogels were advantageous for quick gelation and healthy cell and tissue.

Wound regeneration may be improved by tissue engineering that combines exogenous stem cells with hydrogel scaffolds. However, the current scaffold materials for skin tissue creation frequently do not provide a sufficient microenvironment. Cai and co-workers [44] showed how an enzyme-crosslinked hyaluronic acid hydrogel may support acute wound healing by giving exogenous bone marrow mesenchymal stem cells a growing microenvironment. Dopamine was grafted onto hyaluronic acid to make this substance, which then underwent enzyme crosslinking using hydrogen peroxide and horseradish peroxidase to produce a loose, porous structure. With its sticky and self-healing qualities, the hydrogel provided exogenous BMSCs with a milieu that was highly compatible with cells. According to in vivo research, this hydrogel considerably sped up the healing of acute full-thickness skin lesions, causing little scarring and the development of appendages like hair follicles.

This work offered a viable therapeutic approach for attaining scar-minimized wound healing in addition to a unique skin tissue engineering scaffold.

3.1.3.4 Crystallization-Based Crosslinking

The crystallites that are already there or that form in the polymer chain serve as the foundation for the physical crosslinking of the network throughout this process, which ultimately leads to the creation of a hydrogel. To create a hydrogel, the PVA solution is repeatedly frozen and thawed. Several parameters, including molecular weight, solution concentration, freezing duration and temperature, number of cycles, etc., affect the final gel's characteristics [45].

3.1.4 Types of Hydrogels Based on Degradability

Based on how they break down, hydrogels have been classified as either biodegradable or non-biodegradable. Complete biodegradability is a feature of hydrogels made from natural polymers like fibrin, agarose, alginate, and chitosan. By breaking physical or covalent crosslinks, degradable gels composed of water-soluble polymers, including PEG, PAM, PVA and polyvinylpyrrolidone (PVP) break down. For many biomedical applications, the main requirement for using a material is its biodegradability. N-isopropylacrylamide-stat-acylhydrazide and pectin aldehyde provide the basis of a biodegradable hydrogel that was recently developed as a medication carrier for anticancer treatment, according to Heng et al. [46]. Hydrogels that are not biodegradable, on the other hand, are made using artificial polymers. Although these hydrogels are not ecologically friendly, they have been extensively studied for the development of drug release systems, dental problems, and wound healing [47].

Hydrogels that degrade naturally are now necessary for a wide range of biological applications. This is accomplished by creating biodegradable hydrogels by introducing labile bonds (such as carbonate, amide, ester, carbamate, etc.) into the crosslinks or the polymer backbone. Under physiological settings, these bonds tend to hydrolyze chemically or enzymatically. Crosslinkers that are substrates for both general and particular MMPs (matrix metalloproteases) are also used in formulations to regulate degradation and the kinds of cells that proliferate into the matrices. Through the combination of varying concentrations of hydrolytically degradable crosslinkers, these formulations also provide the capability of further engineering the breakdown rate [48].

Controlling the hydrogels' degradation kinetics for certain uses is quite interesting since it might aid in developing a production procedure. Because of the complexity of the environment, a variety of factors affect the hydrogel's degradation phenomenon after implantation. The hydrogel's stability and performance are also dependent on its biocompatibility and inflammatory nature, as well as the breakdown products it produces.

3.1.5 Classification Related to the Hydrogel's Structure

One important aspect in the classification of hydrogels is their configuration [49]. Based on their chemical structure and physical characteristics, hydrogels are categorized as: amorphous, or non-particle; a complex combination of amorphous and crystalline phases is called semi-crystalline; and sometimes crystalline. Polymers crystallize into layers called lamellae, which are chain-folded layers. Spherulites, or spherical components, can be formed by the clustering of these layers. When the lamellar chains are ordered in predictable patterns, polymers have crystalline phases; when the lamellae are formed irregularly, they have amorphous phases. Usually, polymers will form spherulitic crystals after melting. The nucleation sites, the crystal mass of the polymer, and the cooling percentage all affect the spherulites' diameter. The sequence in the molecular organization of polymers is quantified by the crystallinity degree (%). It can be semi-crystalline (about 95%) or completely amorphous (almost 0%).

3.1.6 Smart Hydrogels According to their Physical Characteristics and Responsivity/Sensitivity

Based on their physical characteristics, the hydrogels are divided into two classes: smart hydrogels and conventional hydrogels. The former has more advanced characteristics than the latter. Changes in their temperature, electric field, pH, enzymatic activities, and light lead the smart hydrogels to react to changes in their external stimuli by changing their network structure, mechanical strength, degradation, swelling performance, etc. These smart hydrogels have a variety of therapeutic uses, including bone tissue engineering, biosensors, and controlled drug delivery. The smart hydrogels are further categorized based on how they respond to stimuli. For example, (i) pH-sensitive hydrogels absorb basic or acidic groups and react to pH changes by gaining or losing a proton. Ionizable functional groups are present in these types of hydrogels. These kinds of hydrogels have ionizable groups, and the binding, association, and dissociation of these ions result in the hydrogel swelling in a solvent. (ii) Thermogels, also known as temperature-sensitive hydrogels, exhibit temperature sensitivity. Because of the establishment of hydrogel linkages, such hydrogels, which contains alkyl groups (hydrophobic), exhibit significant swelling in water-based solutions. Based on how the gels swell and contract as the temperature rises, respectively, the hydrogels that are responsive to temperature are furthermore divided into positive response and negative response gels [50]. Additionally, the way that water molecules interact with the polymer chain's hydrophilic and hydrophobic network changes as the network's solubility varies with temperature. This temperature-dependent network modification results in a sol-gel phase transition, which further classifies the polymer as having UCST-type phase performance and an LCST [51, 52]. Below the LCST, the polymers typically stay in solution, while beyond the LCST, they become hydrophobic and precipitate. Similarly, UCST-type polymers usually convey a reversal change, from gel to sol, but LCST-type polymers may cause a sol-gel transition with heating [53]. (iii) Smart hydrogels that react to an electric field are known as electro-sensitive hydrogels. Based on how the hydrogels react to swelling and deswelling in an electric field, they are divided into dielectric and ionic polymers. Since ionic polymers are primarily conductive, they may release or take charge through the redox process, which allows ions to be delivered to the polymeric network during oxidation reaction and returned to the solution when reduction occurs [54]. On the other hand, by creating Coulombic or electrostatic forces between the electrodes in the polymeric networks, distortion is grounded by the dielectric polymers. Applications for this kind of hydrogel in sensors and actuators have a lot of promise [55]. (iv) By adding certain light-responsive functional groups (chromophores), photo-responsive soft gels can vary their network structure and conformation in response to light and dark. Spiropyran and azobenzenes are the most often utilized chromophores in hydrogels, mostly for controlled drug release applications [56]. (v) Hydrogels sensitive to enzymes are essentially bio-sensitive, that, during cross-linking, include enzymes in their network. Kinases, endonucleases, phosphatases, and proteases are the enzymes that are most needed for this function.

As a result, the description based on the various hydrogel classes can give researchers and scientists a concise guide to choosing hydrogels that meet their needs (Table 3.1). Since the chapter focuses on hydrogel classifications, the following description is predicated on conveying a simple essence of various ferrogels and their functional behavior before delving into the current advancements of these composites.

3.2 Classification of Ferrogels

Extracellular matrix, or ECM, and cellular components, such as various tissue-specific cells, make up the cellular microenvironment. Different forces of varied magnitude are perceived and produced by the cells, including forces at the organism level such as gravity and subcellular forces like actomyosin contractility. These extrinsic (shear, tensile, compressive) and intrinsic (cytoskeletal element formation, reassembly) mechanical stresses have a major influence on cellular biochemical activities. Furthermore, it has also been revealed that the mechanobiological control of the morphogenesis, development, and differentiation of the developing embryo occurs [57]. Therefore, there has been a lot of global interest in the creation of bio-actuator platforms to research how cells react to mechanical stimuli. Bio-actuators are soft actuating platforms that respond mechanically to a variety of stimuli, including light,

Table 3.1 Table summarizing the classification of hydrogels

Classification basis	Categories	Examples
Preparation	Homopolymeric hydrogels	Hydrogels formed from a single monomer species
	Copolymeric hydrogels	Hydrogels composed of two or more types of monomers
	Interpenetrating networks (IPNs)	Networks of two or more cross-linked polymers
Source	Natural hydrogels	Hydrogel derivatives from natural sources like polysaccharides (chitosan, alginate) and proteins (collagen, gelatin)
	Synthetic hydrogels	Hydrogels synthesized from synthetic polymers
	Hybrid hydrogels	Hydrogels blending synthetic and natural polymers
Crosslinking method	Physically crosslinked hydrogels	Hydrogels where crosslinking occurs through physical interactions like hydrogen bonds or hydrophobic interactions
	Chemically crosslinked hydrogels	Hydrogels where crosslinking occurs through covalent bonds
Stimuli response	Thermo-responsive hydrogels	Hydrogels that have the ability to change their properties (e.g., swelling) in response to temperature changes
	pH-responsive hydrogels	Hydrogels that alter their properties in response to pH changes
	Sol-gel hydrogels	Hydrogels that alter their characteristics as the solvent is changed
Ionic charge	Non-ionic hydrogels	Hydrogels with no net charge
	Anionic hydrogels	Hydrogels with a negative charge
	Cationic hydrogels	Hydrogels with a positive charge
Physical aspect	Smart	Smart hydrogels can alter their properties (like swelling or mechanical strength) in response to environmental factors, such as light, pH, or temperature
	Conventional	The only minor changes to conventional hydrogels include low mechanical strength and swelling brought on by external environmental variables
Structure	Amorphous	When the lamellae are structured erratically in hydrogels
	Semi-crystalline	The lamellar chains are arranged in well-known patterns

(continued)

Table 3.1 (continued)

Classification basis	Categories	Examples
Degradability	Biodegradable	Biodegradable hydrogels are biocompatible and can be designed to degrade at a controlled rate
	Non-biodegradable	These hydrogels are designed to maintain their structure and properties for extended periods, without significant degradation. Non-degradable hydrogels are used in applications where long-term stability is required, such as in contact lenses, implants, and some types of drug delivery systems

magnetic fields, pH, temperature, electricity, and so on [58, 59]. Due to its capacity to respond to stimuli by adjusting its water content, hydrogel has emerged as one of the most promising options for creating bio-actuator platforms. Magneto-responsive hydrogels, or "Ferrogels," are one type of stimuli-responsive hydrogel-based bio-actuator that has drawn particular interest from the biomedical community due to its noncontact method of operation. Ferrogels are hydrogels containing a distributed magnetic component found in nature. Transition metal particle oxides such as Fe, Co, Ni, different ferrites, etc., are examples of magnetic characteristics [60, 61]. Magnetostriction, which results from variable polymer networks being bent or their molecular conformation altering, is the capacity of ferrogel to alter its shape or size by an external magnetic field, reliant upon the magnetization behavior of the magnetic particles [62]. Ferrogels are useful in many areas of biomedicine, such as tissue engineering, 3D bioprinting, biosensing, in vitro cellular model production, and many more, by adjusting the blend of magnetically active materials and polymer properties. Mechanotransduction is a critical component of cellular biochemical reactions that it may be able to be remotely regulated. Furthermore, ferrogels can be classified based on interactions and dispersion of magnetic particles.

3.2.1 Ferrogel Types Based on the Interaction of Magnetic Particles

Kickelbick classified ferrogels according to the mechanism of interaction between the polymeric network and the metallic nanoparticles (Fig. 3.4a) [63]. *Class I materials*: Weak physical interactions (hydrogen bonds), weak electrostatic interactions, or van der Waals forces are present in certain ferrogel types (Fig. 3.4b). This class includes the vast majority of ferrogels, which vary widely in their composition of polymers and nanoparticles [64–66]. The magnetite (Fe_3O_4) nanoparticle-embedded polyvinyl alcohol hydrogel is an illustration of this kind of ferrogel [67]. Because the hydrogel contains a hydroxyl group, this polymer exhibits hydrogen bonding [68].

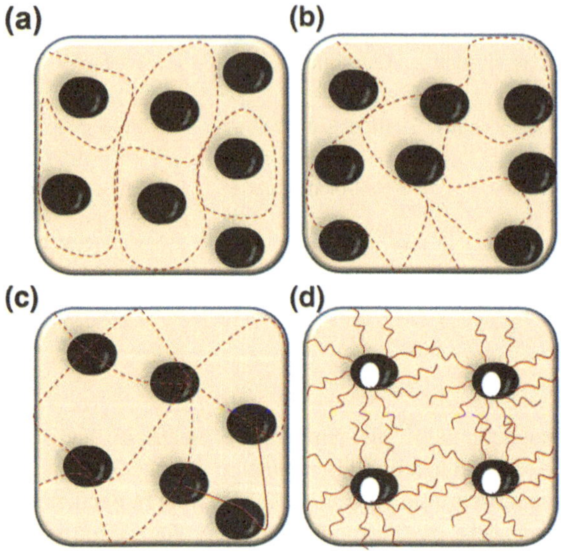

Fig. 3.4 Illustration of the ferrogel framework includes the following: **a** chemically cross-linked polymer with particles enclosed in it; **b** chemically cross-linked polymer of class I that interacts physically with the embedded particles; **c** cross-linked polymer of class II that interacts chemically; and **d** bridged micelles loaded with magnetic particles. Reproduced with permission from Ref. [76]. Copyright 2021 Springer Nature

Materials in class II: These exhibit durable chemical bonds, such as covalent bonds, strong electrostatic interactions, or strong hydrogen bonds, conversely to class I materials (Fig. 3.4c) [69, 70]. Instead of acting as a barrier between the force strengths, this category of materials endures in a constant transition state. Polyacrylamide and polysaccharide in combination with cobalt ferrite ($CoFe_2O_4$) magnetic nanoparticles are the main polymers utilized in this category.

Micelles of a polymer containing magnetic particles: These kinds of ferrogels can be chemically or physically cross-linked polymers having a strong physical bond between the polymeric matrix and the magnetic particles [71, 72]. Magnetic particles are encapsulated in micelles in this form of ferrogel (Fig. 3.4d) [73, 74]. For example, to create these kinds of composite materials, pluronic copolymers—ABA-type triblock copolymers of poly(ethylene oxide) and poly(propylene oxide) are used [75].

Due to the particles' greater flexibility, ferrogels only show minimal interaction with the matrix, which results in particle loss and some toxicity as well as diminished magnetic characteristics. Therefore, for improved performance, the magnetic particles in the solution must be optimized and stabilized. By accumulating the particles, the network of magnetic nodes can prevent aggregation, particle loss, and diffusion, resulting in an increase in the polymer's mechanical strength. Thus, a greater understanding of the various ferrogel types is necessary to select the ideal hybrid material for real-world applications.

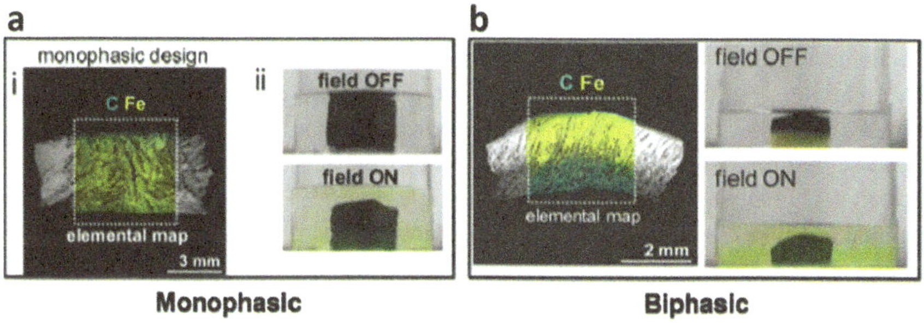

Fig. 3.5 **a** Elemental mapping of a monophasic ferrogel system; **b** elemental mapping of a biphasic ferrogel system to illustrate the distribution of magnetic particles. Reproduced with permission from Ref. [77]. Copyright 2018 Elsevier

3.2.2 *Classification of Ferrogels Based on Dispersion of Magnetic Particles*

Monophasic and biphasic ferrogels are differentiated by the dispersion of magnetic particles within the hydrogel matrix. By evenly dispersing magnetic hydrogel across the hydrogel matrix, monophasic ferrogels are created (Fig. 3.5a). As seen in Fig. 3.5b, on the other hand, the biphasic ferrogels are made by applying an external magnetic field in place while the hydrogel network is forming. A magnetic particle concentration gradient may be produced by applying an external magnet to increase the concentration of magnetic particles on one side of the ferrogel relative to the opposite plane. According to reports, these biphasic hydrogels have better deformability for smaller constructions and increased payload delivery [77]. Ferrogels are categorized as either isotropic or anisotropic by a different team of researchers. Ferrogels with randomly dispersed magnetic particles are said to be isotropic. Anisotropic ferrogels are made when chain-like magnetic particle alignment is accomplished with the application of an external magnetic field [78].

A magnetic self-healing ferrogel that employs the Kabachnik-Fields reaction between poly(vinyl) alcohol (PVA) and phenylboronic acid (PBA) was also described in a current study by He et al. [79]. According to their findings, the dynamic borate ester linkages are responsible for the self-healing process. Therefore, by adjusting the material's properties, ferrogels of various types can be created based on the required application and manufacturing process.

Summary

With the discovery of various polymer structures, crosslinking strategies, processes, characterization procedures, and application fields, hydrogel and ferrogel chemistry has advanced to a new level. This has given rise to new opportunities to find answers for significant biological applications, such as the treatment of cancer, diabetes, bone repair, and cardiovascular issues. The process of getting these gels from the laboratory to the clinical trial and ultimately to the commercial market remains challenging.

Future biomedical research is anticipated to heavily rely on biomedical systems or devices based on hydrogel because of their biodegradability, biocompatibility, and tunability. Additionally, there is potential to create novel hydrogels based on biopolymers (like proteins and polysaccharides), standardize a hydrogel production process, optimize conditions, and apply them to specific biomedical applications like wound healing, tissue engineering, drug delivery, and many more. From this chapter, it can be concluded that there are various types of HG-FGs based on various parameters. The classification of ferrogels, illustrated in this chapter, was a thorough study of the combination of literature. We hope that this chapter will give readers an inclusive outlook of the types of hydrogels and ferrogels in one place.

References

1. N.A. Peppas, J.Z. Hilt, A. Khademhosseini, R. Langer, Hydrogels in biology and medicine: from molecular principles to bionanotechnology. Adv. Mater. **18**, 1345–1360 (2006). https://doi.org/10.1002/adma.200501612
2. N. Zoratto, P. Matricardi, Semi-IPNs and IPN-based hydrogels, in *Polymeric Gels*, (Elsevier, 2018), pp. 91–124. https://doi.org/10.1016/B978-0-08-102179-8.00004-1
3. L. Pescosolido, T. Vermonden, J. Malda, R. Censi, W.J.A. Dhert, F. Alhaique, W.E. Hennink, P. Matricardi, In situ forming IPN hydrogels of calcium alginate and dextran-HEMA for biomedical applications. Acta Biomater. **7**, 1627–1633 (2011). https://doi.org/10.1016/j.actbio.2010.11.040
4. Y.J. Kim, J. Min, Property modulation of the alginate-based hydrogel via semi-interpenetrating polymer network (semi-IPN) with poly(vinyl alcohol). Int. J. Biol. Macromol. **193**, 1068–1077 (2021). https://doi.org/10.1016/j.ijbiomac.2021.11.069
5. L. Wang, H.J. Zhang, X. Liu, Y. Liu, X. Zhu, X. Liu, X. You, A physically cross-linked sodium alginate–Gelatin hydrogel with high mechanical strength. ACS Appl. Polym. Mater. **3**, 3197–3205 (2021). https://doi.org/10.1021/acsapm.1c00404
6. S. Awasthi, S.K. Pandey, H.J. Shwetha, Nehal, S. Selvaraj, Tough, durable and strongly bonded self-healing cartilage-mimicking noncovalently assembled hydrogel nanostructures: the interplay between experiment and theory. Nanoscale **17**, 1616–1643 (2025). https://doi.org/10.1039/D4NR03322A
7. S.W. Ali, S.A.R. Zaidi, Synthesis of copolymeric acrylamide/potassium acrylate hydrogels blended with poly(vinyl alcohol): effect of crosslinking and the amount of poly(vinyl alcohol) on swelling behavior. J. Appl. Polym. Sci. **98**, 1927–1931 (2005). https://doi.org/10.1002/app.22354
8. K.Y. Lee, D.J. Mooney, Alginate: properties and biomedical applications. Prog. Polym. Sci. **37**, 106–126 (2012). https://doi.org/10.1016/j.progpolymsci.2011.06.003
9. L. Almany, D. Seliktar, Biosynthetic hydrogel scaffolds made from fibrinogen and polyethylene glycol for 3D cell cultures. Biomaterials **26**, 2467–2477 (2005). https://doi.org/10.1016/j.biomaterials.2004.06.047
10. I. Mironi-Harpaz, D.Y. Wang, S. Venkatraman, D. Seliktar, Photopolymerization of cell-encapsulating hydrogels: crosslinking efficiency versus cytotoxicity. Acta Biomater. **8**, 1838–1848 (2012). https://doi.org/10.1016/j.actbio.2011.12.034
11. R.V. Ulijn, Enzyme-responsive materials: a new class of smart biomaterials. J. Mater. Chem. **16**, 2217 (2006). https://doi.org/10.1039/b601776m
12. Y. Jia, J. Li, Molecular assembly of Schiff Base interactions: construction and application. Chem. Rev. **115**, 1597–1621 (2015). https://doi.org/10.1021/cr400559g

13. M. van Dijk, D.T.S. Rijkers, R.M.J. Liskamp, C.F. van Nostrum, W.E. Hennink, Synthesis and applications of biomedical and pharmaceutical polymers via click chemistry methodologies. Bioconjug. Chem. **20**, 2001–2016 (2009). https://doi.org/10.1021/bc900087a
14. J. Kalia, R.T. Raines, Hydrolytic stability of Hydrazones and oximes. Angew. Chem. Int. Ed. **47**, 7523–7526 (2008). https://doi.org/10.1002/anie.200802651
15. M.P. Lutolf, J.A. Hubbell, Synthesis and physicochemical characterization of end-linked poly(ethylene glycol)-co-peptide hydrogels formed by Michael-type addition. Biomacromolecules **4**, 713–722 (2003). https://doi.org/10.1021/bm025744e
16. O. Jeon, C. Powell, L.D. Solorio, M.D. Krebs, E. Alsberg, Affinity-based growth factor delivery using biodegradable, photocrosslinked heparin-alginate hydrogels. J. Control. Release **154**, 258–266 (2011). https://doi.org/10.1016/j.jconrel.2011.06.027
17. J.M. Dodda, K. Deshmukh, D. Bezuidenhout, Y.-C. Yeh, Hydrogels: Definition, history, classifications, formation, constitutive characteristics, and applications, in *Multicomponent Hydrogels: Smart Materials for Biomedical Applications*, ed. by J.M. Dodda, K. Deshmukh, D. Bezuidenhout, (The Royal Society of Chemistry, 2023). https://doi.org/10.1039/9781837670055
18. M.J. Moura, H. Faneca, M.P. Lima, M.H. Gil, M.M. Figueiredo, In situ forming chitosan hydrogels prepared via ionic/covalent co-cross-linking. Biomacromolecules **12**, 3275–3284 (2011). https://doi.org/10.1021/bm200731x
19. P. Gacesa, Alginates. Carbohydr. Polym. **8**, 161–182 (1988). https://doi.org/10.1016/0144-8617(88)90001-X
20. W.R. Gombotz, S.F. Wee, Protein release from alginate matrices. Adv. Drug Deliv. Rev. **64**, 194–205 (2012). https://doi.org/10.1016/j.addr.2012.09.007
21. F. Chen, Y. He, Z. Li, B. Xu, Q. Ye, X. Li, Z. Ma, W. Song, Y. Zhang, A novel tunable, highly biocompatible and injectable DNA-chitosan hybrid hydrogel fabricated by electrostatic interaction between chitosan and DNA backbone. Int. J. Pharm. **606**, 120938 (2021). https://doi.org/10.1016/j.ijpharm.2021.120938
22. J. Wang, J. Wei, F. Zhang, X. Huang, X. Dong, H. Jiang, H. Zhang, Y. Bai, K. Xu, P. Wang, Salt-tolerant and self-healing hydrogel containing cationic hyperbranched poly (amido amine) as electrostatic and chemical cross-linking junctions. Polymer (Guildf) **319**, 128063 (2025). https://doi.org/10.1016/j.polymer.2025.128063
23. L. Lei, R. Zhang, R.-X. Bi, Z.-H. Peng, X. Liu, T.-Y. Shi, L. Zhang, R.-P. Liang, J.-D. Qiu, Calcium phytate cross-linked polysaccharide hydrogels for selective removal of U(VI) from tailings wastewater. Water Res. **278**, 123343 (2025). https://doi.org/10.1016/j.watres.2025.123343
24. L.J. Hobson, W.J. Feast, Poly(amidoamine) hyperbranched systems: Synthesis, structure and characterization. Polymer (Guildf) **40**, 1279–1297 (1999). https://doi.org/10.1016/S0032-3861(98)00268-7
25. L.J. Hobson, A.M. Kenwright, A simple 'one pot' route to the hyperbranched analogues of Tomalia's poly(amidoamine) dendrimers. Chem. Commun., 1877 (1997). https://doi.org/10.1039/a704023g
26. P. Li, K. Xu, Y. Tan, C. Lu, Y. Li, P. Wang, A novel fabrication method of temperature-responsive poly(acrylamide) composite hydrogel with high mechanical strength. Polymer (Guildf) **54**, 5830–5838 (2013). https://doi.org/10.1016/j.polymer.2013.08.019
27. K. Xu, X. Liang, P. Li, Y. Deng, X. Pei, Y. Tan, K. Zhai, P. Wang, Tough, stretchable chemically cross-linked hydrogel using core—Shell polymer microspheres as cross-linking junctions. Polymer (Guildf) **118**, 58–67 (2017). https://doi.org/10.1016/j.polymer.2017.04.055
28. P. Li, J. Zhu, P. Sunintaboon, F.W. Harris, New route to amphiphilic core−shell polymer nanospheres: graft copolymerization of methyl methacrylate from water-soluble polymer chains containing amino groups. Langmuir **18**, 8641–8646 (2002). https://doi.org/10.1021/la0261343
29. K. Xu, H. An, C. Lu, Y. Tan, P. Li, P. Wang, Facile fabrication method of hydrophobic-associating cross-linking hydrogel with outstanding mechanical performance and self-healing property in the absence of surfactants. Polymer (Guildf) **54**, 5665–5672 (2013). https://doi.org/10.1016/j.polymer.2013.07.079

30. D.J. Lundberg, J.E. Glass, R.R. Eley, Viscoelastic behavior among HEUR thickeners. J. Rheol. (N Y N Y) **35**, 1255–1274 (1991). https://doi.org/10.1122/1.550174
31. C.M. Hassan, N.A. Peppas, *Biopolymers PVA Hydrogels, Anionic Polymerisation Nanocomposites* (Springer, Berlin, Heidelberg, 2000). https://doi.org/10.1007/3-540-46414-X
32. L. Klouda, A.G. Mikos, Thermoresponsive hydrogels in biomedical applications. Eur. J. Pharm. Biopharm. **68**, 34–45 (2008). https://doi.org/10.1016/j.ejpb.2007.02.025
33. M.A. Ward, T.K. Georgiou, Thermoresponsive polymers for biomedical applications. Polymers (Basel) **3**, 1215–1242 (2011). https://doi.org/10.3390/polym3031215
34. S. Ashraf, H.-K. Park, H. Park, S.-H. Lee, Snapshot of phase transition in thermoresponsive hydrogel PNIPAM: role in drug delivery and tissue engineering. Macromol. Res. **24**, 297–304 (2016). https://doi.org/10.1007/s13233-016-4052-2
35. D.C. Tuncaboylu, M. Sari, W. Oppermann, O. Okay, Tough and self-healing hydrogels formed via hydrophobic interactions. Macromolecules **44**, 4997–5005 (2011). https://doi.org/10.1021/ma200579v
36. Y. Deng, I. Hussain, M. Kang, K. Li, F. Yao, S. Liu, G. Fu, Self-recoverable and mechanical-reinforced hydrogel based on hydrophobic interaction with self-healable and conductive properties. Chem. Eng. J. **353**, 900–910 (2018). https://doi.org/10.1016/j.cej.2018.07.187
37. J. Cao, G. Fan, B. Lu, Exploring the self-assembly mechanisms of wheat gluten polypeptide hydrogels: synthesis and characterization. J. Food Eng. **387**, 112351 (2025). https://doi.org/10.1016/j.jfoodeng.2024.112351
38. M.K. McHale, L.A. Setton, A. Chilkoti, Synthesis and in vitro evaluation of enzymatically cross-linked elastin-like polypeptide gels for cartilaginous tissue repair. Tissue Eng. **11**, 1768–1779 (2005). https://doi.org/10.1089/ten.2005.11.1768
39. L.-S. Wang, C. Du, W.S. Toh, A.C.A. Wan, S.J. Gao, M. Kurisawa, Modulation of chondrocyte functions and stiffness-dependent cartilage repair using an injectable enzymatically crosslinked hydrogel with tunable mechanical properties. Biomaterials **35**, 2207–2217 (2014). https://doi.org/10.1016/j.biomaterials.2013.11.070
40. L. Wang, B. Xu, Y. Nong, P. Wang, Y. Yu, C. Deng, J. Yuan, Q. Wang, Laccase-mediated construction of flexible double-network hydrogels based on silk fibroin and tyramine-modified hyaluronic acid. Int. J. Biol. Macromol. **160**, 795–805 (2020). https://doi.org/10.1016/j.ijbiomac.2020.05.258
41. R. Jin, L.S. Moreira Teixeira, P.J. Dijkstra, M. Karperien, C.A. van Blitterswijk, Z.Y. Zhong, J. Feijen, Injectable chitosan-based hydrogels for cartilage tissue engineering. Biomaterials **30**, 2544–2551 (2009). https://doi.org/10.1016/j.biomaterials.2009.01.020
42. S. Kobayashi, H. Uyama, S. Kimura, Enzymatic polymerization. Chem. Rev. **101**, 3793–3818 (2001). https://doi.org/10.1021/cr990121l
43. B. Bi, H. Liu, W. Kang, R. Zhuo, X. Jiang, An injectable enzymatically crosslinked tyramine-modified carboxymethyl chitin hydrogel for biomedical applications. Colloids Surf. B Biointerfaces **175**, 614–624 (2019). https://doi.org/10.1016/j.colsurfb.2018.12.029
44. B. Cai, J. Fang, S. Zhou, M. Xie, K. Zhang, J. Li, G. Yin, Enzyme-crosslinked hyaluronic acid hydrogel scaffolds for BMSCs microenvironment and wound healing. Int. J. Biol. Macromol. **295**, 139566 (2025). https://doi.org/10.1016/j.ijbiomac.2025.139566
45. C.M. Hassan, N.A. Peppas, Structure and morphology of freeze/thawed PVA hydrogels. Macromolecules **33**, 2472–2479 (2000). https://doi.org/10.1021/ma9907587
46. H. An, Y. Yang, Z. Zhou, Y. Bo, Y. Wang, Y. He, D. Wang, J. Qin, Pectin-based injectable and biodegradable self-healing hydrogels for enhanced synergistic anticancer therapy. Acta Biomater. **131**, 149–161 (2021). https://doi.org/10.1016/j.actbio.2021.06.029
47. E.Y.X. Loh, M.B. Fauzi, M.H. Ng, P.Y. Ng, S.F. Ng, M.C.I. Mohd Amin, Insight into delivery of dermal fibroblast by non-biodegradable bacterial nanocellulose composite hydrogel on wound healing. Int. J. Biol. Macromol. **159**, 497–509 (2020). https://doi.org/10.1016/j.ijbiomac.2020.05.011
48. C. Chokoza, C.A. Gustafsson, K.P. Goetsch, P. Zilla, N. Thierfelder, F. Pisano, M. Mura, M. Gnecchi, D. Bezuidenhout, N.H. Davies, Tuning tissue ingrowth into proangiogenic hydrogels via dual modality degradation. ACS Biomater. Sci. Eng. **5**, 5430–5438 (2019). https://doi.org/10.1021/acsbiomaterials.9b01220

49. S. Awasthi, S.K. Pandey, Recent advances in smart hydrogels and carbonaceous nanoallotropes composites. Appl. Mater. Today **36**, 102058 (2024). https://doi.org/10.1016/j.apmt.2024.102058
50. H.J. Van der Linden, J. Westerweel, Temperature-sensitive hydrogels, in *Encyclopedia of Microfluidics and Nanofluidics*, 2nd edn., (Springer, New York, 2014), pp. 1–5
51. M.R. Matanović, J. Kristl, P.A. Grabnar, Thermoresponsive polymers: Insights into decisive hydrogel characteristics, mechanisms of gelation, and promising biomedical applications. Int. J. Pharm. **472**, 262–275 (2014)
52. J. Bergueiro, M. Calderõn, Thermoresponsive nanodevices in biomedical applications. Macromol. Biosci. **15**, 183–199 (2015)
53. M.T. Cook, P. Haddow, S.B. Kirton, W.J. McAuley, Polymers exhibiting lower critical solution temperatures as a route to Thermoreversible Gelators for healthcare. Adv. Funct. Mater. **31** (2021)
54. O.K. Simya, P. Radhakrishnan, A. Ashok, K. Kavitha, R. Althaf, Engineered nanomaterials for energy applications, in *Handbook of Nanomaterials for Industrial Applications*, 1st edn., (Elsevier, Amsterdam, 2018), pp. 751–767
55. B.N.M. Truong, K.K. Ahn, Modeling and control of hysteresis for DEAP actuator. Sens. Actuators A Phys. **201**, 193–206 (2013)
56. B.P. Bastakoti, Z. Liu, Multifunctional polymeric micelles as therapeutic nanostructures: targeting, imaging, and triggered release. Nanostruct. Cancer Ther., 261–283 (2017)
57. K.H. Vining, D.J. Mooney, Mechanical forces direct stem cell behaviour in development and regeneration. Nat. Rev. Mol. Cell Biol. **18**, 728–742 (2017). https://doi.org/10.1038/nrm.2017.108
58. Q. Shi, H. Liu, D. Tang, Y. Li, X. Li, F. Xu, Bioactuators based on stimulus-responsive hydrogels and their emerging biomedical applications. NPG Asia Mater. **11**, 64 (2019). https://doi.org/10.1038/s41427-019-0165-3
59. D. Quashie, P. Benhal, Z. Chen, Z. Wang, X. Mu, X. Song, T. Jiang, Y. Zhong, U.K. Cheang, J. Ali, Magnetic bio-hybrid micro actuators. Nanoscale **14**, 4364–4379 (2022). https://doi.org/10.1039/D2NR00152G
60. J. Garcia-Torres, Hybrid hydrogels with stimuli-responsive properties to electric and magnetic fields, in Hydrogels - From Tradition to Innovative Platforms with Multiple Applications (IntechOpen, 2023). https://doi.org/10.5772/intechopen.102436
61. W.F. Ibeaho, M. Chen, J. Shi, C. Chen, Z. Duan, C. Wang, Y. Xie, Z. Chen, Multifunctional magnetic hydrogels fabricated by iron oxide nanoparticles mediated radical polymerization. ACS Appl. Polym. Mater. **4**, 4373–4381 (2022). https://doi.org/10.1021/acsapm.2c00359
62. A.P. Safronov, A.Y. Zubarev, E.A. Mikhnevich, E.V. Rusinova, A kinetic model for magnetostriction of a ferrogel with physical networking. Philos. Trans. R. Soc. A Math. Phys. Eng. Sci. **379**, 20200315 (2021). https://doi.org/10.1098/rsta.2020.0315
63. G. Kickelbick, *Hybrid Materials: Synthesis, Characterization, and Applications* (Wiley, New York, 2007)
64. J. Landers, L. Roeder, S. Salamon, A.M. Schmidt, H. Wende, Particle-matrix interaction in cross-linked PAAm-hydrogels analyzed by Mössbauer spectroscopy. J. Phys. Chem. C **119**, 20642–20648 (2015)
65. C. Mayer, V. Cabuil, T. Lalot, R. Thouvenot, Magnetic nanoparticles trapped in pH 7 hydrogels as a tool to characterize the properties of the polymeric network. Adv. Mater. **12**, 417–420 (2000)
66. M. Zrinyi, Intelligent polymer gels controlled by magnetic fields. Colloid Polym. Sci. **278**, 98–103 (2000)
67. R. Ramanujan, L. Lao, The mechanical behavior of smart magnet-hydrogel composites. Smart Mater. Struct. **15**, 952 (2006)
68. F. Yokoyama, I. Masada, K. Shimamura, T. Ikawa, K. Monobe, Morphology and structure of highly elastic poly (vinyl alcohol) hydrogel prepared by repeated freezing-and-melting. Colloid Polym. Sci. **264**, 595–601 (1986)
69. M. Bonini, S. Lenz, R. Giorgi, P. Baglioni, Nanomagnetic sponges for the cleaning of works of art. Langmuir **23**, 8681–8685 (2007)

70. R. Barbucci, D. Pasqui, G. Giani, M. De Cagna, M. Fini, R. Giardino, A. Atrei, A novel strategy for engineering hydrogels with ferromagnetic nanoparticles as crosslinkers of the polymer chains. Potential applications as a targeted drug delivery system. Soft Matter **7**, 5558 (2011). https://doi.org/10.1039/c1sm05174a
71. J.A. Galicia, O. Sandre, F. Cousin, D. Guemghar, C. a Ménager, V. Cabuil, Designing magnetic composite materials using aqueous magnetic fluids. J. Phys. Condens. Matter **15**, S1379 (2003)
72. N. Frickel, R. Messing, A.M. Schmidt, Magneto-mechanical coupling in $CoFe_2O_4$-linked PAAm ferrohydrogels. J. Mater. Chem. **21**, 8466–8474 (2011)
73. S. Li, V.T. John, G.C. Irvin, G.L. Rachakonda, S.H. McPherson, C.J.O. Connor, Synthesis and magnetic properties of a novel ferrite organogel. J. Appl. Phys. **85**, 5965–5967 (1999)
74. M. Babincová, D. Leszczynska, P. Sourivong, P. Čičmanec, P. Babinec, Superparamagnetic gel as a novel material for electromagnetically induced hyperthermia. J. Magn. Magn. Mater. **225**, 109–112 (2001)
75. Q. Jian, I. Asempah, S. Laurent, A. Fornara, R.N. Muller, M. Muhammed, Injectable superparamagnetic ferrogels for controlled release of hydrophobic drugs. Adv. Mater. **21**, 1354–1357 (2009)
76. S. Awasthi, A review on hydrogels and Ferrogels for biomedical applications. JOM **73**, 2440–2451 (2021). https://doi.org/10.1007/s11837-021-04734-7
77. S. Kennedy, C. Roco, A. Déléris, P. Spoerri, C. Cezar, J. Weaver, H. Vandenburgh, D. Mooney, Improved magnetic regulation of delivery profiles from ferrogels. Biomaterials **161**, 179–189 (2018). https://doi.org/10.1016/j.biomaterials.2018.01.049
78. A. Attaran, J. Brummund, T. Wallmersperger, Development of a continuum model for ferrogels. J. Intell. Mater. Syst. Struct. **28**, 1358–1375 (2017). https://doi.org/10.1177/1045389X16672564
79. X. He, Y. Zeng, G. Liu, Y. Tian, Y. Wei, L. Zhao, L. Yang, L. Tao, Magnetic self-healing hydrogel from difunctional polymers prepared via the Kabachnik–fields reaction. ACS Macro Lett. **11**, 39–45 (2022). https://doi.org/10.1021/acsmacrolett.1c00720

Chapter 4
Characterization Techniques and Their Efficacy for Hydrogels and Ferrogels

4.1 Fundamental Characterization Techniques and Their Utilization for HG-FGs

Engineering processes and employments include a wide range of materials that require characterization to identify their chemical, mechanical, tribological, magnetic, biological, and physical characteristics. The capacity to get structural and functional evidence on components across several length scales and dimensions is essential to the development of gel synthesis and characterization. Both novice and experienced scientists and engineers must revisit the fundamentals of fabrication and characterization techniques to have a better knowledge of the approach for a certain application [1, 2]. A sufficient understanding of HG-FG's characterization is necessary for practical applicability. Correlating the performance behavior of materials with their characterization and manufacturing parameters can be facilitated by the meaningful understanding of materials.

Through a variety of potential methods, distinct characterization procedures are used. Transmission electron microscopy (TEM), atomic force microscopy (AFM), scanning electron microscopy (SEM), and other techniques provide access to the materials' surface morphology. Investigations into the materials' phase (X-ray diffraction) and functional groups (Fourier transform infrared spectroscopy) come after the top surface characterizations. Mechanical strength, wear resistance (tribology), magnetism, corrosion resistance, and biological findings result from the application-based characterizations. All these methods for HG-FGs characterization are made possible by the knowledge and assistance of physicists, chemists, material scientists, and instrumentation professionals. There are still gaps in the literature about the explanations of the fundamentals of characterization procedures of hydrogels and ferrogels. This chapter is only state-of-the-art in terms of characterization principles of HG-FGs with a significant discussion about the theoretical aspects of the gel's mechanical and morphological characteristics. This chapter aims to provide a comprehensive overview for both expert as well as beginner researchers working in

S. Awasthi, *Smart Hydrogels and Ferrogels*, Materials Horizons: From Nature to Nanomaterials, https://doi.org/10.1007/978-981-95-9077-3_4

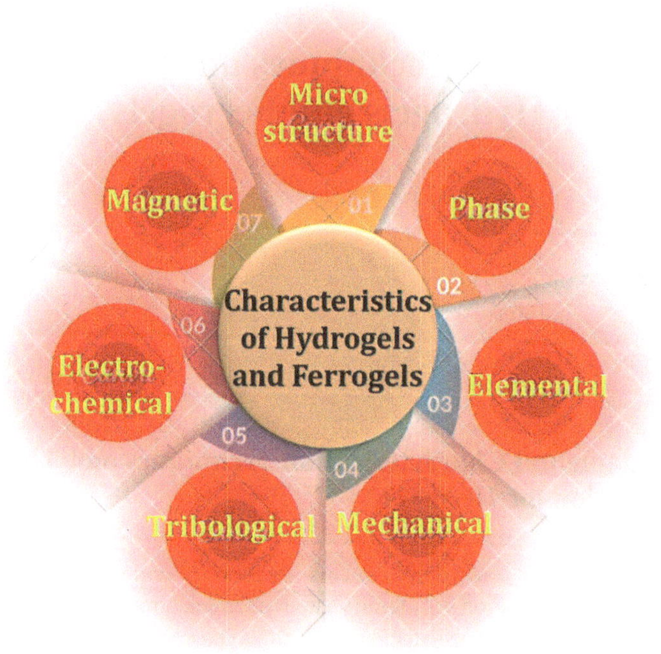

Fig. 4.1 Diagrammatic illustration of the significant advanced methods used in HG-FG's characterization

the field of materials science and engineering, chemistry, physics and biology. A few essential advanced procedures for the materials' manufacture and characterization are summarized in Fig. 4.1. The graphic draws attention to the many characterization methods, such as elemental analysis, mechanical, wear performance (tribology), electrochemical, magnetic, microstructure, and phase characterizations. The next sections give the working principles' interpretation of the aforementioned practices.

4.1.1 Physico-Chemical Properties of Hydrogels and Ferrogels

4.1.1.1 Top Surface Morphology Characterizations

SEM is a versatile instrument for solid object imaging and identifying their constituent elements. The lowest magnification range of SEMs is typically between 20 and 50×, with the largest magnification being mostly constrained by the electron beam's size, which can be up to one million (106) times greater. Elements with sizes ranging from millimeters to nanometers can be measured using these magnification settings. High-end scanning electron microscopes have a sensitivity of about 0.5 nm. The secondary electron image (SE picture), which maps secondary electron emissions according to geographical locations, is the main source of SEM data. Because of the electron beam's perpendicular incidence across the area under study, the surface angle, which correlates to the topography of the sample, affects the quantity of secondary electrons released. The second most prevalent kind of data gathered in SEM imaging is backscattered electrons (BSE), which illustrate backscattered electron emissions according to geographical locations. The secondary electron image (SE picture), which maps secondary electron emissions according to spatial placements to reflect topography, is the main source of SEM data. Because of the perpendicular incidence of the electron beam over the inspection zone, the surface angle, and the topography of the sample affect the quantity of secondary electrons released. In SEM images where electron emission is linked to geographic location, sample roughness cannot often be evaluated from a top-down viewpoint because electron emission may not be correlated with height. Sometimes, even in cases where electron emission does vary with height, such as in secondary electron pictures, it is not possible to accurately determine the height from electron emission data. The sample must typically be seen cross-sectionally in order to evaluate the roughness. SEM documents the morphology of the sample and creates and collects distinct X-rays using energy-dispersive X-ray spectroscopy (EDS). EDS illuminates the elements within the sample and produces element distribution maps by spatially mapping X-ray data. Component concentrations as low as 0.5% by weight may often be detected by EDS within the X-ray production chamber [3].

Transmission electron microscopy (TEM), which has a very high spatial resolution, is an amazing instrument for examining a range of material properties. At the atomic level, we know the size distribution, crystal structure, shape, strain, defects, and chemical information. All of the information gathered by TEM is provided by the interaction of electrons with the substance. The term "transmission" electron microscopy comes from the fact that images are created by identifying electrons that have flowed through a thin object. The composition of the sample and the specific information needed for TEM characterization dictate that TEM materials must be exceptionally thin, typically with a thickness of less than 200 nm to allow for effective electron transmission [3]. There are several methods used in transmission electron microscopy, and each one gives the materials unique information. Here are a few instances of these techniques.

High-Resolution TEM (HRTEM): This method uses the interference between transmitted and diffracted beams to produce images at the atomic level. For viewing atom configurations in the projection, HRTEM is quite useful.

Energy-Filtered TEM (EFTEM): The elements and compounds are mapped using this method. It may also be used to improve contrast in images and diffraction patterns by reducing inelastically distributed electrons.

Energy-Dispersive X-Ray Spectroscopy (EDS): This technique is helpful for determining a specimen's surface chemical composition. It makes it easier to characterize the chemical or compositional makeup of materials.

Selected Areas Electron Diffraction (SAED): This is utilized to provide optimal circumstances for dark field imaging, define crystal morphologies, and ascertain the direction of nanowire formation.

Besides SEM and TEM, a flexible piece of equipment called a scanning probe microscope (SPM) can take precise height images of the surfaces of solid samples and offer details on their mechanical, electrical, and magnetic properties. A small physical probe is positioned near the sample to measure its responses as it moves laterally over the surface. A range of probe techniques, including scanning tunneling microscopy and magnetic force microscopy, are included in the general phrase "scanning probe microscopy." Atomic force microscopy (AFM) frequently uses surface imaging to investigate features including form, roughness, or mechanical characteristics. Silicon (Si) is used to make the great majority of commercially available probes. Typically, these probes consist of a cantilever beam with a sharp tip constructed of etched silicon. The sort of probe utilized determines the type of data collected. For general-purpose height imaging, a standard etched Si cantilever or tip is frequently utilized. When specific data, such magnetic information, is required, a magnetized tip may be created by applying a magnetic substance on a standard tip [4]. A vertical cantilever oscillation is used in the tapping mode of operation, which is rather prevalent. The tip "taps" the sample by getting near enough to make contact with it during the bottom phase of the oscillation. A feedback loop is used to keep the oscillations' amplitude steady and the force between the tip and the sample constant. In atomic force microscopy (AFM), the most common kind of data produced is a height map. The term "height" map may be a bit deceptive since it basically tracks how the feedback loop moves the probe to keep the force between the tip and the sample constant. The field of view of AFM pictures is usually less than 10 mm, while the greatest scan size is around 100 mm. These images may detect atomic sizes. AFM frequently requires ultrahigh-vacuum (UHV) conditions to attain atomic resolution. The sharpness and form of the physical probe have an impact on the resolution, and rather than magnification, the field of vision defines AFM pictures. Since AFM images provide absolute height information, sample roughness may be easily determined. The tip's physical characteristics, however, may have an impact on data accuracy by making wells appear smaller and protrusions larger than they actually are. The tip itself can occasionally be imaged by sharp sample features, which is usually obvious and suggests likely tip damage. The disposable AFM tips will wear out during imaging. The required data type and the sample's modulus often specify the tip and cantilever characteristics [5].

The SEM, TEM, and AFM techniques are potentially applied to observe special features of hydrogels and ferrogels. For instance, Yang et al. [6] got an idea from the strong cavity-wall structure of the cuttlebone with significant energy absorption and develop a robust hierarchical predesigned hydrogel assembly that integrally synchronously combine various organic and inorganic micro-nano building blocks to diversified morphologies. Using hydrogen, covalent bonds, and electrostatic interactions, the two types of predesigned hydrogels are built layer by layer into stiff micro-hollow structures and brick-and-mortar structures, respectively (inspired by cuttlebones). As a result of natural evolutionary selection, hierarchical networking is a fundamental optimization technique for healthcare systems that increases functional efficiency while minimizing resource consumption [7]. Scanning electron microscopy (SEM) images and scanning electron microscopy/energy-dispersive X-ray spectrometry (SEM/EDS) images were taken using a GeminiSEM 450 Schottky field emission scanning electron microscope at acceleration voltages ranging from 1 to 5 kV. Each sample was gold-sputtered for 45 s at a constant current of 30 mA before being observed. In particular, once the original solvent had been sufficiently replaced with ethanol in a critical point drier, the initial hydrogel was dried using supercritical CO_2 for the SEM sample preparation of various beginning hydrogels.

The crosslinked cCNF created a three-dimensional nanonetwork in the wall layer hydrogel, and the amino mica was evenly distributed throughout this network (Fig. 4.2A a, b). The carboxylated cellulose nanofiber (cCNF) securely enveloped the amino mica, which piled consistently to produce a sturdy wall layer in the brick-and-mortar construction (Fig. 4.2A c). The quaternized cellulose nanofiber (qCNF) and PVA create nanonetworks in the stiff cavity layer hydrogel, supporting the even dispersion of GB (Fig. 4.2A d, e). In the meanwhile, surface B-O-C bonds allow the glass bubble (GB) to interact with polyvinyl alcohol (PVA). The material eventually reaches a close-packed form as the space between GB progressively shrinks throughout the dehydration and densification process (Fig. 4.2A d). Tightly packed and shielded by the PVA/qCNF combination, the GB forms a close-packed, stiff micro-hollow structure with walls that are roughly tangent to one another (Fig. 4.2A f). The advantages of assembling predesigned hydrogels in a single step are further shown by the densified cuttlebone-inspired structural materials' lack of a discernible gap between the rigid cavity and wall layers (Fig. 4.2A g). Additionally, a partly incorporated dovetail-like structure that contributes to increased damage tolerance is shown at the interface between the two layers (Fig. 4.2A h). From the outside to the inside, a clear and sturdy, stiff cavity-wall organized microstructure resembling a cuttlebone can be seen by 3D reconstruction (Fig. 4.2A i).

Magnetic hydrogels (ferrogels) and their top surface characterization have been the subject of several studies. In one of the latest studies, the microstructural analysis was done on the ferrogels that are made from agar gum and octenyl succinic anhydride (OSA) starch coated with a stabilizing agent (PNiPAAm). A thorough TEM evaluation was carried out to shed light on the size, shape, and structural flaws of colloidal iron nanoparticles (Fig. 4.2B). As previously noted for magnetic nanoparticles (MNPs) coated with polymers, Fig. 4.2B a illustrates the occurrence of aggregates (clustering) of MNPs in Agarose pores. As will be covered in more detail in the rheology section, this clustering appears to be self-limiting in the gels since it does not grow larger than tiny clusters and does not result in phase separation. TEM images at low magnification (Fig. 4.2B b) show that the samples are made up of faceted MNPs. They also showed that MNPs are structurally homogeneous. Additionally, it was observed that there were three populations of various diameters:

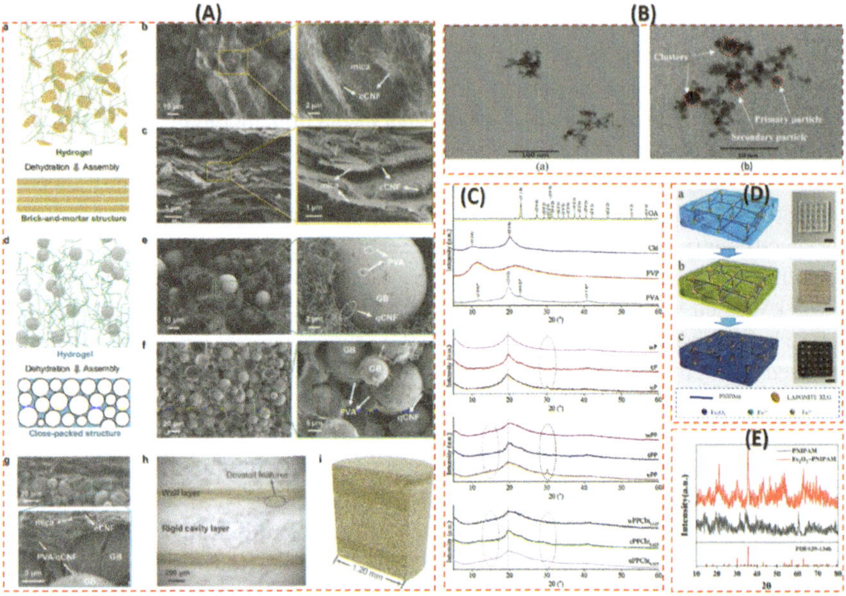

Fig. 4.2 **A a** Diagram showing the construction of a brick-and-mortar wall from its initial hydrogel to its thick state. Yellow platelets are amino mica; green curves are cCNF. **b** SEM pictures of the first hydrogel's asymmetrical mica platelet arrangement. **c** SEM pictures show the RCWSM's physical barrier. **d** Diagram showing the stiff cavity layer's transition from the original hydrogel to the dense condition. qCNF green curve, PVA blue curve, and GB grey sphere. **e** SEM pictures show the GB's asymmetrical distribution in the original hydrogel. **f** SEM pictures of the rigid cavity-wall structural material's (RCWSM) stiff cavity layer. **g** SEM pictures of the contact between the tightly packed stiff cavity layer and the brick-and-mortar wall. **h** A polarizing microscope view of the RCWSM's cross-section demonstrates the clear dovetail features at the two-layer interface. **i** The RCWSM was reconstructed in three dimensions using X-ray microtomography. Reproduced with permission from Ref. [6]. Copyright 2025 Springer Nature (Open access under the license CC-BY-NC-ND). **B** Ferrogel TEM micrographs at various magnifications. Reproduced with permission from Ref. [28]. Copyright 2025 Wiley & Sons. **C** X-ray diffractograms of all hydrogel formulations, crosslinkers, and polymers. Reproduced with permission from Ref. [10]. Copyright 2025 Royal Society of Chemistry (Open access under the license CC-BY-NC-ND). **D** Magnetic gel's 3D printing method. **a** Temperature-sensitive PNIPAM hydrogel that was 3D printed. **b** Iron ion solution-soaked PNIPAM hydrogel. **c** After soaking in NaOH solution, a magnetic hydrogel was developed. **E** XRD analysis of Fe_2O_3 + PNIPAM magnetic hydrogels and PNIPAM temperature-sensitive hydrogels. Reproduced with permission from Ref. [26]. Copyright 2025 MDPI (Open access under the license CC-BY-NC-ND)

(i) particles that are primary and have average sizes between 3 and 4 nm, (ii) particles that are secondary and have average diameters between 6 and 9 nm, and (iii) clusters that are a combination of primary and secondary particles with average diameters between 12 and 13 nm.

4.1.1.2 Phase Analysis

XRD is an extremely flexible technique that provides chemical information for elemental and phase research. XRD is highly significant for chemical characterization, textural analysis, and stress measurements. Although XRD measurement may identify the degree of crystallinity in a polymer, it requires crystalline samples. The use of XRD for thin-film research is now feasible because of new optical techniques, even though it has typically been used to evaluate bulk materials. The Bragg diffraction law is used in the procedure. Additionally, the XRD method works very well with ceramic and polymeric materials [8, 9]. X-ray diffraction serves as the primary method for determining the type of bonding and the arrangement of crystalline structure within an amorphous polymeric nanocomposite. Because polymeric nanocomposite materials develop crystalline behavior after synthesis, they perform exceptionally well in X-ray diffraction. XRD can identify changes in the microstructure and interlayer spacing of the sample. Poly(vinyl alcohol) polymeric samples and graphene oxide were produced and reported in the literature. Additionally, XRD was used to characterize the composites. The various peaks vanished when reduced graphene oxide and graphene oxide were added to the polymeric matrix, leaving just one peak discernible in XRD. The results of the research demonstrated enhanced crystallinity as graphene oxide and decreased graphene oxide accumulation in the bulk polymer.

Himawan and team [10] examined the properties of the dried PVA-based hydrogels and their constituent parts in the solid state using an X-ray diffractometer (XRD). For this aim, an XRD apparatus with a Cu Kβ radiation source was used, running at 40 kV and 25 mA of current. Each specimen was placed without any indentation on a zero-background holder. The continuous PSD quick scan mode was used to scan the samples in the 2θ range of 5–60° at a rate of 6° min^{-1}. Diffrac.Eva® software was then used to process and analyze the diffractograms. The diffractogram shown in Fig. 4.2C showed many peaks of both monoclinic and orthorhombic PVA. The orthorhombic lattice in the (2 0 2) plane was represented by the strong, wide peak at 19.7°. The monoclinic lattice in the (2 0 0), (0 0 2), and (3 1 4) orientations is responsible for the last three halo peaks at 11.6°, 22.9°, and 40.7°, respectively [11, 12]. All these peaks, nevertheless, were quite wide, suggesting that the virgin PVA polymer employed in this investigation had a low level of crystallinity [13]. The amorphous form of the poly(vinylpyrrolidone) (PVP) polymer was represented by the usual halo peaks in the PVP diffractogram, which were located at around 11.3 and 21.8° 2θ [14]. When acetyl groups are removed from the polymer backbone, intermolecular hydrogen bonds are broken, causing these peaks to become wider and less strong in chitosan [15, 16]. The α form of the anhydrous crystal was the kind of oxalic acid (OA) employed in this investigation, according to the OA diffractogram [17]. According to data obtained from the software, this form was distinguished by prominent main peaks at 22.9, 31.0, and 37.4° 2θ, which corresponded to the (1 1 1), (2 0 1), and (0 2 2) planes of the orthorhombic lattice of the material [17].

The diffraction patterns of all hydrogel film compositions were comparable. Every sample had a distinctive halo peak at 19.5°, which is attributed to the orthorhombic

PVA's (2 0 2) plane. All crosslinked formulations (cP, cPP, and cPPChi0.625) displayed stronger peaks at this diffraction angle, according to the data, which may have indicated a higher degree of crystallinity. This characteristic may be explained by the crosslinking process's elimination of any remaining water and the heating process's abundant energy introduction [18–22]. The removal of the formulation's peaks at 14.8 and 30.3° upon crosslinking is another noteworthy aspect. The (2 0 0) and (− 1 1 1) planes of the OA dihydrate form (not included in the graph) may be responsible for these peaks [23]. It's possible that OA recrystallized when the polymeric solution dried. The peak disappearance could not be ascribed to either melting or decomposition since the thermal analysis findings showed that none of these processes occurred at the temperature utilized for the crosslinking method. No further OA peaks were seen following the crosslinking procedure, which might be explained by the fact that OA was consumed by the crosslinking reaction. For PVA hydrogels crosslinked with suberic acid, it has also been documented that the crosslinker peak in the formulation vanishes upon crosslinking [24].

To get a deeper comprehension of the structure and characteristics of ferrogels, XRD (X-ray diffraction) analysis is utilized to examine the crystalline structure, composition, and physical characteristics of the magnetic nanoparticles (MNPs) contained inside the gel matrix. Basak et al. [25] fabricated gelatin-based ferrogel with high magnetism iron oxide (magnetite, IO) particles coated with polyethylene glycol (PEG). Powder-XRD investigations were used to further characterize the PEGylated IO particles and identify their phases. The crystal planes (220), (311), (400), (422), (511), and (440) of the cubic crystal structure were confirmed by the JCPDS file no. 01-075-0033, and prominent peaks at 2θ of 30.4, 35.6, 43.4, 53.5, 57.2, and 62.8° were reported by commercial and synthesized IO particles. The most prominent peak at 35.6° corresponds to the (311) plane. Furthermore, the IO_Coated (Low) and IO_Coated (High) diffractograms showed a polymeric hump with PEG-20000 peaks at 2θ of 19 and 23.3°. The IO_Coated (High) sample showed an increased-intensity hump at 2θ of 17–27°, confirming the magnetite phase of the synthesized particles and the increased intensity of the polymeric hump of PEG-20000 as the reaction mixture's PEG concentration increased.

Furthermore, ferrogels made from poly-N-isopropylacrylamide (PNIPAM) that are 3D printed were successfully manufactured in a bilayer using 4D printing technology by Li and co-workers [26]. Figure 4.2D shows the magnetic hydrogel formation process. The process starts with printing PNIPAM hydrogels with a specific microarray assembly utilizing the PNIPAM hydrogel printing procedure with XLG, as explained in the approaches of the current study. The sample appears as a translucent and colorless gel, as shown in Fig. 4.2D a. The printed PNIPAM gel in Fig. 4.2D b is submerged in a deionized water solution that contains iron ions (Fe^{2+}, Fe^{3+}), which causes the sample to become light yellow. After submersion, a NaOH solution is applied to the sample to create a black magnetic hydrogel, as shown in Fig. 4.2D c. The XRD patterns of Fe_2O_3 + PNIPAM magnetic hydrogel and PNIPAM temperature-sensitive hydrogel were shown in Fig. 4.2E a for the 20–80° (2θ) range. The presence of Fe_2O_3 in the hydrogel system was demonstrated by the fact that all of the distinctive diffraction peaks of Fe_2O_3 + PNIPAM magnetic hydrogels were in good agreement with the typical pattern of magnetic Fe_2O_3 particles (JCPDS card No. 39-1346) [27].

4.1.1.3 Functional Group Characterizations

An efficient method for a comprehensive material analysis is infrared (IR) spectroscopy, which offers information on the molecular dynamics, structure, and chemical composition of the materials. Due to its ease of use, broad applicability, time and money efficiency, safety, nontoxicity, and minimal space requirements, FT-IR finds wide-ranging applications. Furthermore, FTIR spectroscopy is an essential technique for describing polymeric and biopolymeric materials. It has been shown that FTIR technology is useful for detecting and describing copolymers, homopolymers, and polymer hybrids. Additionally, it was successfully applied to evaluate the polymerization method and describe the surface, modification, degradation, and structure of the polymer. A change in the absorption region's distinctive form suggests that the material's structure has changed or that pollutants are present. Among the many nondestructive testing methods, FTIR spectroscopy proves to be an exceptionally versatile tool for evaluating the surface characteristics of various materials [29]. Additionally, it makes it possible to track chemical reactions at the material's surface in real-time, which enables a detailed examination of how these methods change with respect to different applications. Furthering solid-state physics is another benefit of FTIR spectroscopy. This method might be used to both chemical analysis and the investigation of infrared absorption by electronic carriers. The spectrometer must be attached to a certain configuration to fully utilize its features [30, 31].

The efficacy of FTIR for hydrogels and ferrogels can be recognized by describing functional groups, probing structural alterations, and examining interactions within the gel network. Rosacea is a persistent inflammatory skin disorder that primarily impacts the face's center and is characterized by persistent erythema and paroxysmal skin flushing. There will be capillary dilatation, papules, and pustules in more severe instances. Furthermore, ocular abnormalities, hyperplasia, and hypertrophy may be present in certain cases [32–35]. Chen et al. [36] confirmed the functional groups in poloxamer 407 (P407) and poloxamer 188 (P188) blended thermos-responsive hydrogels, encapsulating carvedilol. The FTIR spectra of Car-P40724/P1881, P40724/P1881, and carvedilol are displayed in Fig. 4.3A. At around 2882 cm^{-1} (C–H stretching) and 1102 cm^{-1} (C=O stretching), P40724/P1881 displayed distinctive peaks [37, 38]. In accordance with the literature, N–H, O–H, and C=C were identified as the causes of the distinctive peaks at 3343 and 1587 cm^{-1} for pure carvedilol [39]. Interestingly, Car-P40724/P1881 had carvedilol and P40724/P1881 characteristic peaks, indicating that carvedilol was successfully added to the P40724/P1881 hydrogels.

Because of their high conductivity and flexibility, sodium alginate (SA)-based conductive hydrogels offer promising uses in personal medical diagnostics and human activity monitoring. The majority of sensing SA-hydrogels, however, have limited practical uses because of their reduced mechanical qualities and absence of

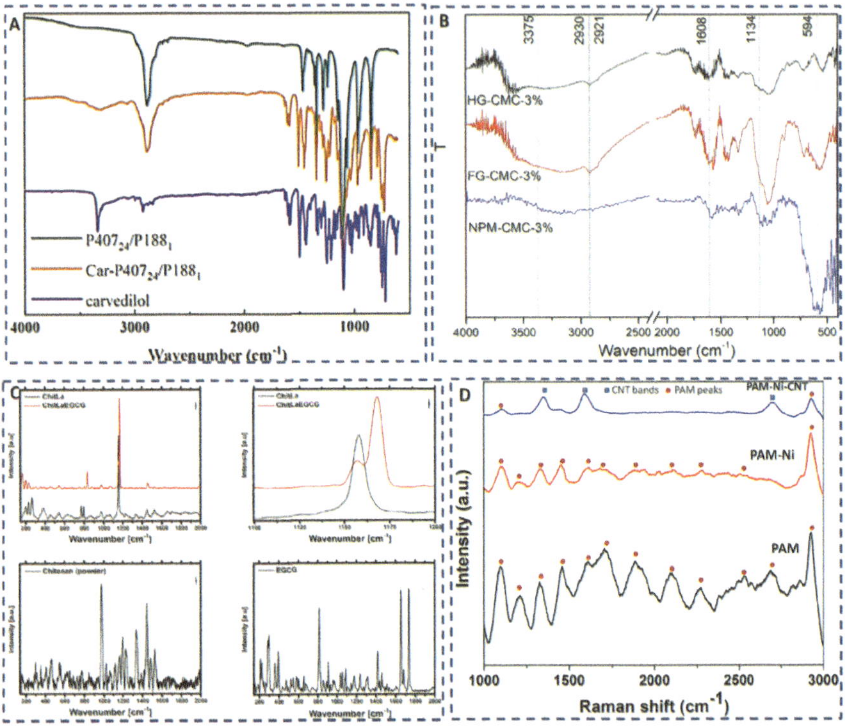

Fig. 4.3 **A** FTIR characterization of P407$_{24}$/P188$_1$ and Car-P407$_{24}$/P188$_1$. Reproduced with permission from Ref. [36]. Copyright 2025 American Chemical Society (Open access under the license CC-BY-NC-ND). **B** Ferrogel (FG-CMC-3%), hydrogel (HG-CMC-3%), and nanoparticle (NPM-CMC-3%) FTIR spectra. Reproduced with permission from Ref. [50]. Copyright 2024 Springer Nature (Open access under the license CC-BY-NC-ND). **C** The material's Raman spectra before and after sorption (**a**), the range 1100–1200 cm^{-1} (**b**). Pure chitosan (**c**) and EGCG (**d**). Reproduced with permission from Ref. [67]. Copyright 2023 Elsevier. **D** PAM, PAM-Ni, and PAM-Ni-CNT hydrogel composites' Raman spectra. Reproduced with permission from Ref. [67]. Copyright 2023 Elsevier

antibacterial, self-healing, and self-adhesive properties. Consequently, a straightforward method was used in the study conducted by Wei et al. [40] to create a PAA-SA hydrogel with a multifunctional double network made of sodium alginate (SA) and poly(acrylic acid) (PAA). FTIR examined the interaction mechanism between the hydrogel. The hydroxyl (–OH) stretching vibration of the AC-Bt surface layer and its interlayer water may be responsible for the distinctive peaks at 3659.0 and 3460 cm^{-1}. The –OH peaks of PAA-SA hydrogel at 3659, 1610, and 913 cm^{-1} vanished when compared to AC-Bt, suggesting that the polymer chain's –COOH interacts contains the hydroxyl groups that are active on AC-Bt, causing AC-Bt to take part in the hydrogel network's crosslinking. Furthermore, the peak of Si–O–Si in PAA-SA hydrogel (antisymmetric stretching vibration) shifted from 1032 to 1066 cm^{-1}, suggesting that the hydrogel network crosslinking altered the chemical environment of the AC-Bt crystal [41]. Additionally, the PAA-SA hydrogel had Bt characteristic peaks around 504 cm^{-1}, indicating once more that AC-Bt played a role in the hydrogel's network of organic and inorganic development. The C–H, –COOH, and –COO– absorption bands on PAA were identified as the source of the distinctive spectral peaks at 1454, 1726, and 1556 cm^{-1} [42]. SA's distinctive peaks, which correspond to the –OH, C–C, and –COO–, respectively, emerged at 3436, 1433, and 1680 cm^{-1} [43, 44]. At 1672 cm^{-1}, the distinctive peak in the PAA-SA hydrogel that was ascribed to –COOH emerged. The layer Ca^{2+} from AC-Bt and the –COO– on SA may create a cross-link. In the PAA-SA hydrogel, the distinctive peak of –OH ascribed to SA was moved to 3419 cm^{-1}. The PAA-SA hydrogel network's extensive –COOH and –OH hydrogen bonding with the PAA and SA chains were revealed by the FTIR data.

According to their varying cross-link densities, Akin and team [45] formulated different types of ferrogels named as GF4, GF8, GF12, and GF16, based on the glutaraldehyde contents of 4, 8, 12, and 16 wt. %, respectively. The FTIR spectra of ferrogels, gelatin hydrogels, raw gelatin, and Fe_3O_4 show a large peak in the 3200–3600 cm^{-1} region in all samples, which is associated with O–H stretching vibrations [46]. Amide I, II, and III bands are the distinct absorption spectral peaks at 1600–1800, 1470–1570, and 1250–1350 cm^{-1} that are indicative of amide groups of gelatin peptide linkages [47]. The spectra for ferrogels, gelatin hydrogels, and raw gelatin all contain these distinctive bands. The C–H stretching is represented by the other distinctive band, which is seen between 2800 and 2950 cm^{-1} [48]. A peak at 537 cm^{-1} in the FTIR spectra of Fe_3O_4 indicates vibrations of the Fe–O bonds [49]. The plot for gelatin ferrogels also shows the distinctive peak.

Medina and the team [50] developed polysaccharide/iron oxide ferrogels and characterized them with SEM, XRD, FTIR, etc. An effective method for learning more about the interaction between iron oxide and polymers is FTIR. As can be seen from Fig. 4.3B's FTIR spectra, the hydroxyl group from carboxymethyl cellulose (CMC)'s stretching vibration is located at about 3375 cm^{-1} and the distinctive C–H stretching of CH_2 from CMC at 2930 cm^{-1}, which validates the combination of CMC and MNPs. Additionally, the bands at 2921 cm^{-1} (stretching) and 1134 cm^{-1} (bending) are associated with the C–H vibrational modes, while the distinctive peaks of CMC at about 3420 (stretching) and 1608 cm^{-1} (bending) are ascribed to the hydroxyl modes, respectively [51]. Additionally, the ferrogel and nanoparticles exhibit a peak at 594 cm^{-1}, which is ascribed to the Fe–O vibration for iron oxides [52].

4.1.1.4 Structural Features

Since the Raman effect was discovered in 1928, Raman spectroscopy has been experiencing a resurgence, which has produced a number of innovative techniques. Research on nanomaterials has expanded in recent decades because of fascinating developments in the fields of optics, quantum mechanics, and spectroscopy. Raman spectroscopy is among the greatest analytical methods for describing the structure of many different kinds of materials, such as ceramics, polymers, carbonaceous allotropes, etc. Raman spectroscopy makes it simple to determine the sp^3 and sp^2 fractions, diameter, chirality, and structural characteristics of graphite, graphene, carbon nanotubes, and diamond-like carbon (DLC), respectively [53–55]. The Raman spectroscopy approach is crucial for characterizing polymer materials since it requires little sample handling to obtain a spectrum. This technique is utilized in several applications include tracking polymerization events, forecasting physical characteristics, and mapping polymeric phases [56]. In the basic principles of Raman spectroscopy, a substrate is exposed to a laser beam with frequency υ_0 and energy E_0 ($E_0 = h\upsilon_0$, where h is the Planck constant). As a consequence, there are faint light reflections with marginally differing frequencies and an elastically dispersed beam from υ_0 (nonelastically scattered light). One of the vibration modes of the material is represented by the frequency shift υ_1 (s^{-1}). When c is the speed of light, the Raman shift (cm^{-1}) equals υ_1/c. The scattered light with $\upsilon_0 - \upsilon_1$ is referred to as Stokes scattering, while that with $\upsilon_0 + \upsilon_1$ is known as anti-Stokes scattering, even if the materials' dispersive effect causes a frequency difference between the Stokes and anti-Stokes sides of a Raman mode. Raman measurements of the materials are usually performed using the Stokes scattering [57, 58].

In situations with a high-water content, where it is insensitive to water molecules, Raman spectroscopy is an important technique for examining hydrogels and offering insights into their chemical structure, composition, and physical characteristics. In a recent study, for the first time, lanthanum-modified chitosan hydrogel was produced. Epigallocatechin gallate (EGCG) was released under regulated conditions using the hydrogel as a carrier. The spectra of the material before and after EGCG modification are displayed in Fig. 4.3C. As can be observed, there are several bands in the spectrum prior to sorption, most of which are produced from chitosan. But some bands can be attributed to $La(OH)_3$. The interactions between La^{3+} and OH^- are the source of three distinct bands with the Raman shifts 201, 229, and 265 cm^{-1} [59]. For example, the two bands for the Raman shift 385 and 454 cm^{-1} are typical bands that may be matched to chitosan. They can be attributed to the pyranoid ring's out-of-plane bending vibrations. Additional chitosan-derived bands may be found at the Raman shifts of 1229 and 1448 cm^{-1} [60]. C–C/C–O stretching is responsible for the first band, whereas C–H bond deformations are responsible for the second [61, 62]. There are also two distinct bands at 767 and 797 cm^{-1}. These bands cannot be attributed to either glycerin or chitosan, according to published studies and the spectra of pure chitosan (Fig. 4.3C) [63]. Therefore, the emergence of these bands may indicate coordination interactions between La^{3+} ions and the functional groups of chitosan. There is also a very strong band with a Raman shift of 1060 cm^{-1}. It is evident that pure chitosan does not exhibit this distinctive band. On the other hand, a band at around 1150 cm^{-1} in pure chitosan can be attributed to the C–N bond's stretching vibrations. We could see this rise in band strength due of the interaction between nitrogen atoms and La^{3+} ions. The spectra showed notable alterations following the EGCG sorption stage.

Due to its potential applications in wearable electronics, field-assisted medication delivery, magnetic biosensors, magnetic field-sensitive transformers, strain sensors, and more, polyacrylamide hydrogel containing nickel nanoparticles (ferrogel) has drawn more and more interest. Wear can destroy the durability and magnetic sensitivity of conventional ferrogels, perhaps rendering them unusable for future applications. The polyacrylamide (PAM) hydrogels were prepared by Awasthi et al. [9] with the encapsulation of carbon nanotubes (CNT) and nickel (Ni). The distinctive bands in Fig. 4.3D verified the existence of CNTs in the PAM-Ni-CNT hydrogel samples. For PAM, the Raman bands correspond to about 1116, 1216, 1330, 1458, and 1610, 1714, 1888, 2100, 2268, and 2925 cm^{-1} [64]. For CNTs, the bands correspond to the D-band, G-band, and 2D band at approximately 1340, 1591 and 2699 cm^{-1} [65]. While the G band is caused by the stretching vibration of sp^2 bound C–C bonds, which are connected to the E_{2g} mode, the D band is linked to various defect-induced and lattice mismatches [66].

Another essential step in protecting the material's surface is to observe its mechanical, tribological, magnetic, and electrochemical strengths. Thus, the HG-FG's tribological, magnetic, electrochemical, and biological characteristics are covered in sequential sections after the microstructure and phase characterization.

4.1.2 Mechanical Characteristics of HG-FGs

A Vickers macro-indenter with many indents, a set load, and a predetermined dwell duration at room temperature may be used to measure the bulk hardness of the materials. Optical microscopy may be used to measure the diagonals of indents, and the following formula can be used to determine hardness (Hv). In the equation, d is the average diagonal length, and P is the applied force. The typical set-point force may be used to map the samples' nanomechanical dynamic modulus.

$$\mathrm{Hv} = 1.854\left(\frac{P}{d^2}\right) \tag{4.1}$$

However, a diamond-tipped nanoindenter (Berkovich indenter with a radius of curvature of around 150 nm) may be used to measure nano hardness and decreased modulus. Figure 4.4A b displays a normal load–displacement diagram throughout the indentation's implementation, whereas Fig. 4.4A a illustrates the operating idea. The penetration depth increases with loading, reaching its maximum depth under excessive load. However, due to elastic recovery, the penetration depth decreases when the force is reduced during unloading. A residual impression that demonstrates the material's plastic deformation is left behind when the indenter is removed (Fig. 4.4A a). The hardness (H) is determined from the penetration depth using the Oliver–Pharr technique (Fig. 4.4A b) [68].

$$H = \frac{P_{\mathrm{max}}}{A_{\mathrm{contact}}} \tag{4.2}$$

The following formula is used in the calculation of elastic modulus.

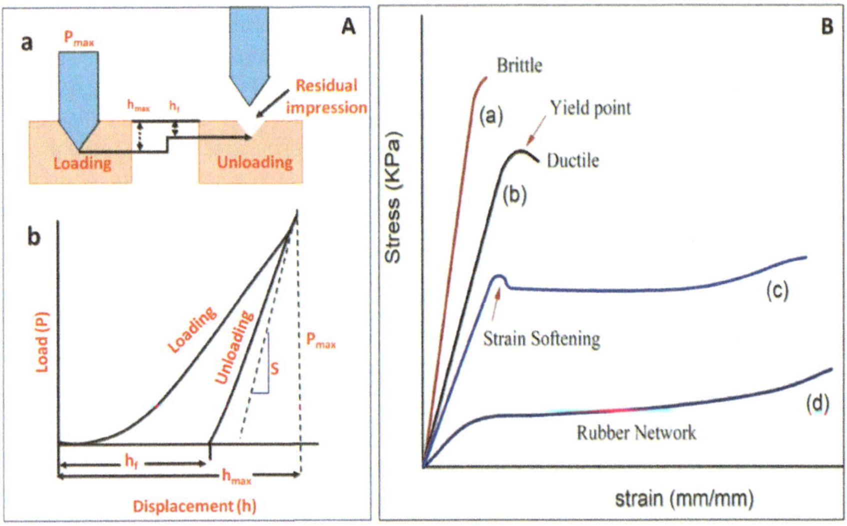

Fig. 4.4 **A** Diagram illustrating **a** the nanoindentation's operation and **b** the loading and unloading load–displacement curve. Reproduced with permission from Ref. [68]. Copyright 2024 American Chemical Society (Open access under the license CC-BY-NC-ND). **B** Numerous stress–strain curves are displayed by polymers, each of which depicts unique characteristics. Glassy polymer behavior is shown by curves (**a**–**c**), whereas the rubbery state is shown by curve (**d**). The critical points of material breakdown are located at the extremities of these curves: **a** showing brittleness, whereas **b**–**d** showing ductility. Put more simply, under stress, various kinds of polymers flex and break differently. Rubbery polymers, represented by curve (**d**), act differently from glassy polymers, as indicated by curves (**a**–**c**). Certain polymers, like glass, break quickly when strained (curve **a**), whereas others stretch and distort before shattering (curves **b**–**d**). Reproduced with permission from Ref. [79]. Copyright 2025 Springer Nature (Open access under the license CC-BY-NC-ND)

$$E_r = \left[\frac{1 - v_i^2}{E_i} - \frac{1 - v_s^2}{E_s}\right]^{-1}, \tag{4.3}$$

where the planned region of contact between the substance and the indenter is denoted by $A_{contact}$ and the maximal indentation load by P_{max}. Nevertheless, v_s and E_s represent the Poisson ratio and elastic modulus of the sample, respectively, and v_i, and E_i represent the Poisson ratio and elastic modulus of the intender [69].

Rubber elasticity theory is commonly used to assess the mechanical properties and structure of polymer gels, which have long been recognized in terms of their similarity to swollen rubbers [70, 71]. According to this idea, gels are networks composed of neighboring crosslinks connected by sub-chains. According to this viewpoint, a polymer network's shear modulus, denoted by G, may be expressed simply as [72]:

$$G = \nu k_B T(\text{affine-network}). \tag{4.4}$$

When T is the absolute temperature, ν is the crosslink density, and k_B is the Boltzmann constant. According to the affine network model, crosslink junctions stay fixed and uniformly flow with the network. However, real networks show variations in these crosslink locations, which reduces the free energy of the system. Several hypotheses, such as the slip-link model, constraint junction model, and the phantom network model, have been put forth to explain these variations [72–74].

For instance, according to Gaussian statistics, the bulk of connections vary, while a tiny percentage remains constant, according to the phantom network concept. As a result, the equation [72, 75] is modified:

$$G = (\upsilon - \mu)k_B T(\text{phantom-network}) \tag{4.5}$$

$$\mu = \frac{2\upsilon}{f}. \tag{4.6}$$

In this case, redefining υ and μ to represent actual networks with flaws is more appropriate: The number density of "effective" crosslinks is shown by μ, while the number density of elastically "effective" strands (subchains between crosslinks) is represented by υ. Shear tests or stress-elongation curves are used to calculate the shear modulus, which is represented as [72]:

$$\sigma = G\left(\lambda - \frac{1}{\lambda^2}\right), \tag{4.7}$$

A commercial rubber ring composed of natural rubber that has been vulcanized in bulk ($\phi = 1.0$) and in an equilibrium swelling condition in toluene at ambient temperature ($\phi \approx 0.18$), as well as their stress-draw ratio graphs, show the differences. Since $\nu = \phi\upsilon 0$, where $\upsilon 0$ and υ are the number densities of effective chains in the bulk and swelled states, respectively, the drop in υ causes the swelling-induced fall in G.

However, because of the complicated network structure and gel production procedures, examining the link between the structure and mechanical properties of polymer gels is more difficult. For instance, copolymerization of crosslinkers and monomers frequently results in nonuniform networks with a variety of topological flaws and spatial irregularities. As a result, swollen rubbers often have better mechanical qualities than ordinary gels. The link between tough gels' structure and mechanical properties is examined in the following sections of the chapter. First, the theoretical basis of scattering for polymer solutions and gels are discussed, and then the structure–mechanical property relationship is examined. By varying the degree of crosslinking or regulating the temperature during production, hydrogels designed for biomedical applications can have their mechanical characteristics precisely adjusted to satisfy particular requirements [76]. These changes make it possible to precisely optimize the stiffness of hydrogels, which is essential for their performance in a variety of applications. It is necessary to conduct thorough studies that take into consideration the material characteristics, experimental setup, and research goals in order to comprehend the complex aspects affecting the mechanical behavior of these polymeric materials. For example, a universal testing machine is frequently used to evaluate the elasticity of hydrogel, but rheometers or dynamic mechanical analysis equipment are usually used to examine viscoelastic characteristics [77]. It is important to remember that the Young Modulus in hydrogels is closely related to the water-gel matrix interaction (Fig. 4.4B). Selecting the right hydrogel stiffness is crucial for a variety of applications, from creating heterogeneous prosthetic devices such as replacements for intervertebral discs to choosing appropriate materials for osteoblast cell versus adipocyte seeding [78].

The compression test is another useful method for assessing the mechanical characteristics of various hydrogel types. To determine mechanical characteristics depending on applied pressure and compression distance, this technique compresses the hydrogel between two plates. Although it needs a flat surface, the compression test provides flexibility in hydrogel geometry unlike extensometer-based testing since it is not restricted to particular forms like rings or strips. However, this method has drawbacks such as hydrogel bulging and unequal pressure distribution. To combat bulging, techniques such as enclosing the hydrogel around its periphery have been investigated; however, this alteration may change the measurement properties. Despite these difficulties, the compression test has been helpful in a number of research projects when examining the mechanical characteristics of cell-seeded hydrogels [74].

Water-soluble, non-toxic, hydrophilic, stable, viscoelastic, and biocompatible are the characteristics of polyacrylamide (PAM) hydrogels. PAM is a synthetic polymer composed of either acrylamide alone or acrylamide plus acrylic acid. PAM may form bonds with other functional groups such as –COOH, $-NH_2$, and –C=O groupings [80] because it contains several main amido groups. However, hydrogels' brittleness and poor mechanical qualities limit their use with cartilage scaffolds [81]. Because the matrix and reinforcing agents have a high interfacial and mechanical interlocking, the reinforcing agents may enhance the material's properties in every manner. Graphene, carbon nanotubes (CNTs), titanium oxide (TiO_2), silicon oxide, and other biocompatible reinforcing agents can be employed to reinforce hydrogels. For the strengthening of PAM hydrogels, Awasthi et al. [82] used a combination of CNT and TiO_2 encapsulation within the PAM matrix. Compression and needle insertion testing were performed for the examination of the mechanical strength of the hydrogels. The hydrogels' ability to withstand punctures is a crucial feature for examining tissue damage from within. The needle insertion approach may be used to study the hydrogels' behavior. This method may be helpful in assessing cartilage damage after surgery or in simulating the condition that results from a sharp instrument inadvertently entering the cartilage and causing injury. Although the needle insertion approach may provide very valuable information for soft materials, it has not yet been properly studied for hydrogels. For the first time, the produced hydrogels' resistance to punctures was described for cartilage applications.

Hypodermic needle insertion studies were conducted using a displacement-controlled drive. A linear variable differential transformer coupled to a linear reciprocation system recorded the needle displacement, and a force sensor with a precision of

1 mN was used to quantify the insertion force [82]. The distance between the needle and the hydrogel surface was 15 mm, and the needle insertion speed was maintained at 0.33 mm/s. The insertion force functions as a function of time (or displacement), according to force versus displacement charts for needle piercing through various gels. Conical fracture creation causes the force to shift in stages, which causes the insertion force on the needle to drop sharply. When pure PAM was used, the total insertion force rose to around 0.16–0.25 N before dropping precipitously to roughly 0.06 N (Fig. 4.5i). The maximum force for the PAM-TiO_2-CNT sample, however, was only ~ 0.035 N after the same insertion, at a depth of ~ 12.5 mm, and stayed at ~ 0.02 N for the remainder of the displacement. This showed that conical fractures are far less common than PAM (7 cracks per 12.5 mm displacement) during the same displacement length. In the case of PAM, a larger insertion force signifies greater stress in the gel surrounding the needle, which leads to increased gel degradation. The resistance offered by the material surface was represented by the first force step. Conical fractures were far less common in PAM-TiO_2-CNT-based samples (1 shallow conical crack in 12.5 mm displacement), which indicates that the sample has the least amount of internal damage, similar to tissue healing. Pure gel reaches its maximum net force, indicating that the gel generates its greatest resistance forces along the needle. This demonstrates that pure PAM is the toughest to pass through, and the material that breaks the most internally after the needle tip is penetrated. When compared to PAM, the conical fractures in the PAM-TiO_2 and PAM-CNT samples were nearly three times smaller.

Cartilage is a high-water-content, load-bearing tissue that has exceptional strength and flexibility. Nevertheless, creating cartilage-adaptive biomaterials to replace or regenerate injured cartilage tissue remains a significant challenge. Zhou and the research team [83] used protein templating to create chitosan nanofibrillar hydrogels (O–CN gels), which have water content and mechanical characteristics comparable to cartilage. High tensile strength and Young's modulus (15.4 MPa and 24.1 MPa, respectively) are demonstrated by the O–CN gels. When the EtOH treatment duration was extended from 6 to 24 h, the tensile strength of double-network chitosan/protein (gelatin or silk fibroin) hydrogels (DN CG) gels enhanced from 2.0 to 3.2 MPa. The gelatin of chitosan–gelatin (CG) gels was mostly responsible for this improvement in mechanical characteristics as it was able to create stronger internal hydrogen bonds with EtOH during the partial dehydration process. The tensile strength of the crystal-mediated random chitosan nanofibrillar hydrogels (R-CN) (after a 24-h EtOH treatment) increased from 8.5 to 11.8 MPa as the NaOH treatment duration increased from 12 to 24 h, but the gels' modulus stayed relatively constant at $\approx$ 4.2 MPa. The R-CN gels' high physical crosslinking density in a high-concentration NaOH solution was the primary source of this reduced mechanical characteristics, which may have an impact on the tangling of chitosan chains and the development of chitosan crystals [84, 85]. The R-CN gels treated with EtOH for 6 and 12 h showed comparable changes in mechanical characteristics. In contrast to gelatin or chitosan–NaOH gels, which were too weak and brittle to support this weight, the produced R-CN gels demonstrated an impressive weight-carrying capability of 1 kg (1 mm width $\times$ 1 mm thickness).

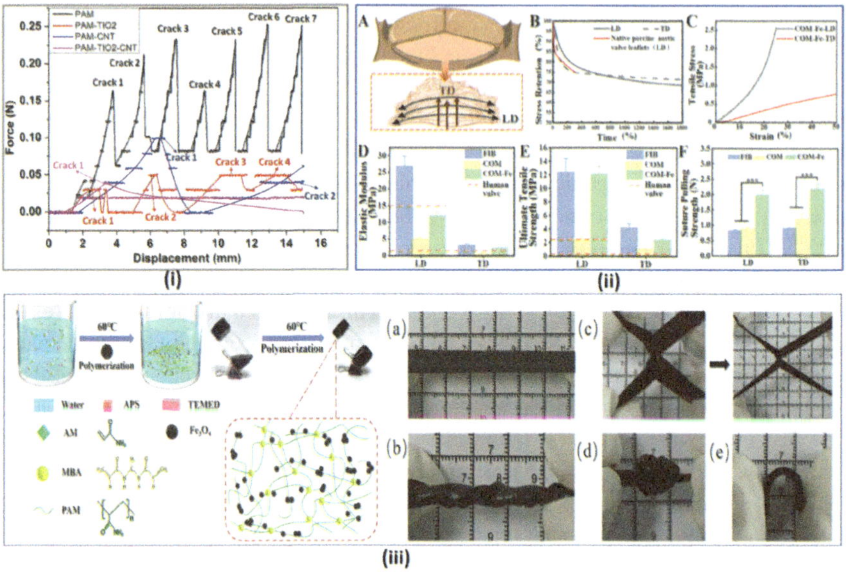

Fig. 4.5 **i** Hydrogel sample force–displacement curves during a needle insertion test. Reproduced with permission from Ref. [82]. Copyright 2021 American Chemical Society. **ii** Comparing the mechanical characteristics of normal tissues with those of FIB, COM, and COM-Fe [20]. **A** A schematic illustration of the native heart valve's composition and fiber dispersion. LD: the leaflet's circumferential direction; TD: its radial direction. The COM-Fe material's stress–strain and stress–relaxation curves for tensile testing in the LD and TD directions are shown in **B** and **C**, respectively. **D** Elastic modulus; **E** FIB, COM, and COM-Fe ultimate tensile strength; and **F** suture pulling strength. Reproduced with permission from Ref. [92]. Copyright 2025 Elsevier. **iii** The mechanical properties of the PAM-Fe_3O_4 magnetic hydrogels include stretching, twisting, cross-stretching, knotting, and bending (left image, **a**–**e**, respectively). Reproduced with permission from Ref. [99]. Copyright 2024 MDPI (Open access under the license CC-BY-NC-ND)

In addition to having outstanding tension characteristics, the R-CN gels displayed a stress of 21.2 MPa at 80% compression strain. The R-CN gels maintained their structure under these compression conditions, but the CG gels broke quickly. At 40% compressive strain and less than 35% load loss, the R-CN gels could sustain 10,000 continuous compression-relaxation cycles. Remarkably, the R-CN gels' compressive modulus rose from 0.9 to 1.8 MPa and their compressive stress improved from 21.2 to 27.6 MPa following 10,000 consecutive compression-relaxation cycles. These findings showed that the chitosan nanofibrillar chains' hydrophobic and hydrogen bonding interactions with NaOH treatment allowed the R-CN gels to attain outstanding mechanical performances. Hydrogels typically have little water retention and excellent mechanical qualities [86–88]. Nonetheless, the resulting R-CN gels maintained a rather high 75 wt. % water content.

Soft electrical devices require conductive hydrogels with reliable sensing capabilities. However, at below-freezing temperatures, these hydrogels have a tendency to harden and sustain structural damage, which can result in material degradation and device failure. The primary problem is the effectively developing the micro/nanostructure to improve mechanical characteristics and reliable strain sensing while avoiding freezing in hydrogels. In recent research, a quick method for creating a strain sensor based on the MXene bridging double-network structure utilizing polyacrylamide and agar hydrogels that can continue to function steadily at – 30 °C, which is exceptionally low was described [89]. The findings of rheological tests showed that H-MPA (hydrogel with MXene/PAM/Agar) was synthesized as an elastic gelled state since its storage modulus was larger than its loss modulus at all frequencies. Furthermore, throughout the reaction step, both catalytic polymerization and hydrogen bonding ought to occur. These results lead to the belief that MXene effectively initiates and gels PAM chains, chemically connecting the PAM and agar double network. The resultant H-MPA exhibited remarkable stretchability (stretched five times) at room temperature by taking advantage of this microstructure. Since the inclusion of MXene, the conductivity of the hydrogel has increased tenfold, which is crucial for its use as a flexible sensing material.

To create programmable hydrogels with locally heterogeneous reinforcement skeletons, or interpenetrating double networks, a sequential photopolymerization process coupled with site-selective patterning exposure was reported [90]. This process was inspired by heterostructures and materials found in nature, such as leaves and insect wings. Strong multiple hydrogen bonds created by phase transition allow the heterogeneous interface between stiff skeletons and soft matrices to interlock easily. Numerous mechanical properties of hydrogels, such as modulus (0.32–5.92 MPa), toughness (0.15–18 kJ m^{-2}), dissipated energy (1–100 kJ m^{-3}), impact resistance, and mechanical anisotropy, can be easily sculpted within a single material system by utilizing the size, shape, and distribution of the patterned stiff skeletons. This eliminated the need for designing and optimizing the intricate and elusive material formulation on demand. This straightforward yet effective technique, which relies on heterogeneous patterning, was thought to direct the creation of functional hydrogel materials with programmable mechanical characteristics toward possible engineering uses like flexible circuits and dampening.

Li et al. [91] created gelatin methacrylate/alginate (G/A) hydrogels with cellulose-composited ionic-covalent entanglement (ICE) networks using microfibrillated and nanofibrillated cellulose (MFC and NFC, respectively). Two different length scales were used to evaluate these hydrogels' mechanical properties. Cellulose improves the ICE hydrogels' toughness, compression resistance, and Young's modulus, as shown by macro-compression. The hydrogen bonding that produces a synergistic interaction between ICE and the cellulose fiber network explains this. Young's moduli of ICE hydrogels are more noticeably improved by pure MFC (363.43 kPa) or NFC (325.75 kPa) alone than by a mixture of the two with different contents (261.5–288.5 kPa), since homogeneous fiber networks enhance the structural integrity of the system. Micro-indentation shows that the hydrogen bonds between cellulose and ICE decrease the time-dependent deformation of the cellulose-composed ICE hydrogel. By mapping the hydrogel's Young's modulus and hardness distribution, three different forms of enhancement were found: cellulose fiber networks, ICE synergistic networks, and cellulose fiber networks with ICE synergies. Creep tests showed that cellulose-composed ICE hydrogels are resistant to creep because of MFC's greater confinement effect. In conclusion, the multiscale characterization demonstrated the mechanical reinforcing processes of the cellulose fibers in ICE hydrogels, suggesting possible methods for improving the mechanical properties of the hydrogels.

An important component of human blood circulation is the heart valve, and the creation of artificial heart valves (AHVs) has emerged as a major area of study for cardiovascular disorders. Its progress has been constrained, nonetheless, by structural variations from natural valves. In the work carried out by Wang et al. [92], polycaprolactone gelatin (PCL-Gel) co-spinning directional nanofibers (FIB) were used to produce a three-layer structure of orientation layer-random layer-orientation layer that mimics natural valves. The PCL-Gel/PAAm-co-PAA-Fe composite (COM-Fe) was made by crosslinking the oriented fiber membrane wrapped in polyacrylamide polyacrylic acid copolymer hydrogel (COM) with iron ions. The COM-Fe material has anisotropy similar to native valves and fully meets the thickness requirements for transcatheter surgery. The COM-Fe material does not substantially lose strength or structure with short curling compression, per the findings of in vitro simulated compression. The distribution of the native valve fiber structure determines the circumferential direction (LD), whereas the vertical direction of the fiber orientation determines the radial direction (TD) (Fig. 4.5ii A). To evaluate the movement of internal molecules under constant tension, the composite stent was put through a stress relaxation test. The stability of the material during several opening and shutting cycles in vivo was postulated based on the material's relaxing performance (Fig. 4.5ii B). In order to maximize the mechanical performance of the composite stent and make it more like natural valves, the scientists used a number of different spinning process settings. After screening many samples, the COM-Fe material showed mechanical properties that were almost identical to those of natural valves. The spinning speed, spinning time, and other parameters were effectively regulated. A tensile strength test was performed on the composite stent, and the stress–strain curves in the LD and TD directions showed that COM-Fe exhibited a "J"-shaped curve similar to that of natural tissue when exposed to tensile action (Fig. 4.5ii C). The LD and TD directions' elastic moduli of 15.29 ± 0.36 MPa and 1.98 ± 0.06 MPa, respectively, were similar to the radial (2.0 ± 1.5 MPa) and circumferential (15.6 ± 6.4 MPa) elastic moduli of the human body's natural aortic valve leaflets (Fig. 4.5ii D, where the dashed orange line represents the human valve). However, the ultimate tensile strengths of COM-Fe in the LD and TD directions were 12.25 ± 1.10 MPa and 2.56 ± 0.11 MPa, respectively, whereas the circumferential direction of the human valve had 2.6 ± 1.2 MPa and 0.42 ± 0.24 MPa (Fig. 4.5ii E) [93]. The results showed that the COM-Fe material's suture tensile strength in both the LD and TD directions was significantly higher than that of the FIB and COM samples when the 6–0 suture was employed to assess the material's suture performance (Fig. 4.5ii F). The composite structure of the COM-Fe material is sufficient to support the suture tensile stress and may be firmly fastened to the valve frame without excessive traction holes at the suture site or structurally weak nodes, thereby effectively resolving the problem of a high AHV reflux rate. By fully using the controlled structural characteristics of polymer materials, the results of the mechanical performance test demonstrated that COM-Fe materials with highly biomimetic mechanical capabilities can be generated continuously. More importantly, this mechanical attribute would likely show better in vivo expression, such as less concentrated stress and an appropriate transvalvular pressure gradient, as real valves do. COM-Fe materials thus have a great opportunity for usage in AHV.

Because of their continuous and real-time monitoring capabilities, implantable flexible mechanical sensors have shown significant promise in health monitoring and illness detection. However, the power supply and connections needed for modern electronics are dangerous and inconvenient. Although magnetic-based devices have shown promise in passive and wireless sensing, their applications in biomechanical monitoring have been constrained by their inadequate sensitivity, poor biocompatibility, and mismatched mechanical characteristics. Based on a gelatin methacrylate (GelMA)/Fe_3O_4 magnetic hydrogel, a flexible magnetic strain sensor that is both wireless and passive has been developed by Zhang and co-workers [94]. The sensor has exceptional magnetic properties, ultrasoft mechanical properties, and long-term stability in saline solution. It can detect stresses as tiny as 50 μm. To choose the hydrogel system for the strain sensor, the mechanical properties of the GelMA hydrogel sheet at concentrations between 5 and 15% were evaluated. As the concentration of GelMA rose, so did Young's modulus. The efficacy of hydrogel cross-linking was confirmed by the storage modulus, which is much higher than the loss modulus. The 5% GelMA hydrogel was selected as the sensor hydrogel system for further sensor development because of its mechanical properties, which are comparable to those of soft tissues [95, 96]. Magnetic GelMA hydrogel sheets with different MNP concentrations (0–15 mg/mL) were then made using the same cross-linking method. The mechanical properties of the magnetic hydrogels increased with the addition of MNPs; the hydrogel containing 15 mg/mL MNPs had a storage modulus of 380 Pa and a Young's modulus of 1.2 kPa. Compared to previous flexible sensor technologies disclosed for cardiomyocyte detection, such as metal meshes (2.59–24.1 MPa) [97] and flexible electrodes (3.39–44.71 kPa), the magnetic hydrogels containing 15 mg/mL MNPs showed a much lower modulus [97], and 1.28 MPa conductive elastomers [98]. The energy storage modulus and loss modulus demonstrated that the extra MNP had no effect on the hydrogel cross-linking. The mechanical properties of the hydrogel films loaded with random MNPs showed that the guided arrangement of MNPs did not substantially change the hydrogel modulus (less than 5%). This work demonstrates the potential of magnetic sensing for biomechanical monitoring and provides recommendations for the design of passive and wireless implanted devices.

Although stimuli-responsive hydrogels have great potential for use in biomedical applications, traditional gelation techniques frequently fall short of the complexity and accuracy needed for sophisticated features like tissue engineering, soft robotics, and targeted drug administration. A class of 3D-printable magnetic hydrogels with changeable stiffness, adhesion, and magnetic responsiveness was described by Xuan et al. [99] using a simple and efficient "one-pot" process (Fig. 4.5iii, left picture). This technique enables fine control over the mechanical characteristics of the hydrogel, such as toughness from 243 to 1400 kJ/m^3, tensile strength from 93 to 421 kPa, and elastic modulus from 43 to 277 kPa, by varying the quantities of acrylamide (AM) and Fe_3O_4 nanoparticles. Under an alternating magnetic field, these hydrogels heat up quickly—they reach 44.4 °C in 600 s at 15% weight—and show promise for treating moderate magnetic hyperthermia. The elasticity and strength of PAM-Fe_3O_4 magnetic hydrogels are superior to those of pure PAM hydrogels. These hydrogels also exhibit high ductility, as shown in Fig. 4.5iii, which enables them to withstand considerable tensile deformation both in their initial condition and upon crossing (Fig. 4.5iii a, c). Additionally, PAM-Fe_3O_4 magnetic hydrogels may bend, twist, and knot (Fig. 4.5iii b, d), among other mechanical properties. Additionally, by including Fe_3O_4 nanoparticles and nanoclay into the AM precursor, the rheological characteristics are optimized, allowing for the creation of intricate, high-fidelity structures using extrusion-based 3D printing. This 3D-printable platform offered improved capabilities for usage in magnetic actuation and hyperthermia in biomedical applications by combining high adhesion, multifunctionality, and tunable mechanical characteristics in a unique way compared to other magnetic hydrogels. This development is a major step toward the scalable manufacturing of smart hydrogels of the next generation for bioengineering and precision medicine.

The above-mentioned studies give a brief overview of the recent advancement in the mechanical properties of HG-FGs. It is now simpler for the researcher to address the most important issues before experimental implementation due to these developments. With the reduction of test site expenses and trial-and-error impacts, the issues associated with the formulation of the best HG-FGs with high mechanical strength can meet the needs of the new applications. Understanding these relationships enables the design of HG-FG composites to be optimized while recognizing quick limits and constraints across a wide range of characteristics or situations. Table 4.1 introduces some other recent studies based on the mechanical behavior of the HG-FGs. The subsequent sections offer a concise evaluation of the tribological properties of successful HG-FG composites. In addition to providing guidance on best practices during the characterization and analysis processes, a brief discussion of the frictional and wear behavior is introduced. These sections emphasize the important areas where the most in-depth study is desired and hope to assist researchers in creating future materials in accordance with society's needs.

4.1.3 Load-Bearing Advancements of HG-FGs

The materials may be worn using reciprocal friction and wear testing (Fig. 4.6A a), which employs a counter body (steel ball), a fixed load, stroke amplitude, and frequency. After the fretting test, the damaged surfaces of the samples may be cleaned of wear particles using ethanol. The extent of damage and wear scars caused by the fretting wear of the samples is then evaluated using an SEM scan. Additionally, energy dispersive spectroscopy and an optical surface profilometer may be used to determine the samples' degree of oxidation and surface roughness (Ra), respectively.

Table 4.1 Various mechanical properties of hydrogels concerning different crosslinking methods, crosslinkers, and polymerization techniques [79]

Types of hydrogels	Crosslinking method	Cross-linkers	Polymerization technique	Toughness (kJ m^{-3})	Mechanical stress (kPa)	Fracture stress (kPa)	References
Poly (2-acrylamido-2-methylpropanesulfonic acid) and polyacrylamide	Chemical	*N*,*N*′-methylenebisacrylamide	Free-radical			1700	[98]
Poly(vinyl alcohol), Polyacrylamide, and polyacrylic acid	Physical		Crystallization			990	[100]
Agar and xanthan gum	Physical (hydrophobic interactions)	$CaSO_4$				3640	[101]
Chitosan and polyacrylamide	Both physical and chemical	Na_2SO_4 and *N*,*N*′-methylenebisacrylamide	Free-radical			3700	[102]
Alg and polyacrylamide	Both physical and chemical	CaSO4 and *N*,*N*′-methylenebisacrylamide	Free-radical			156	[103]
Agar and polyacrylamide	Physical (hydrogen bond)	*N*,*N*′-methylenebisacrylamide	Free-radical			1000	[104]
Polyacrylic acid	Chemical	Graphene oxide-based crosslinker	Free-radical	4078	248		[105]
Polyacrylic acid	Both physical and chemical	BF-Si NPs	Free-radical	1385	157		[106]
Polyacrylamide	Both physical and chemical	BF-Si NPs	Free-radical	655	131		[107]
Polyacrylamide	Both physical and chemical	Graphene oxide-based crosslinker	Free-radical	1775	446		[107]

(continued)

Table 4.1 (continued)

Types of hydrogels	Crosslinking method	Cross-linkers	Polymerization technique	Toughness (kJ m^{-3})	Mechanical stress (kPa)	Fracture stress (kPa)	References
Polyacrylamide and DOPA	Physical	CNC	Free-radical	3210	39		[108]
Polyacrylic acid	Chemical	*N*,*N*′-methylenebisacrylamide and cellulose	Free-radical		1500		[109]

The working mechanism of the optical profilometer is shown in Fig. 4.6A b. Wear volume, number of cycles, stroke length, and load are used to compute wear rate (W_r) [68].

$$\text{Wear rate}\,(W_r) = \frac{\text{wear volume}}{\text{number of cycles} \times \text{stroke length} \times \text{load}} \tag{4.8}$$

Archard's equation is used for the wear constant (K) calculation utilizing load (P), wear rate (W_r), and hardness (H).

$$K = \frac{H \times W_r}{P}. \tag{4.9}$$

The pin-on-disc (PoD) assessment approach is a widely used method for researching material sliding properties and wear processes in several tribological situations. A reliable and accurate tribological characterization instrument for coatings, lubricants, and bulk materials is the pin-on disk tribometer.

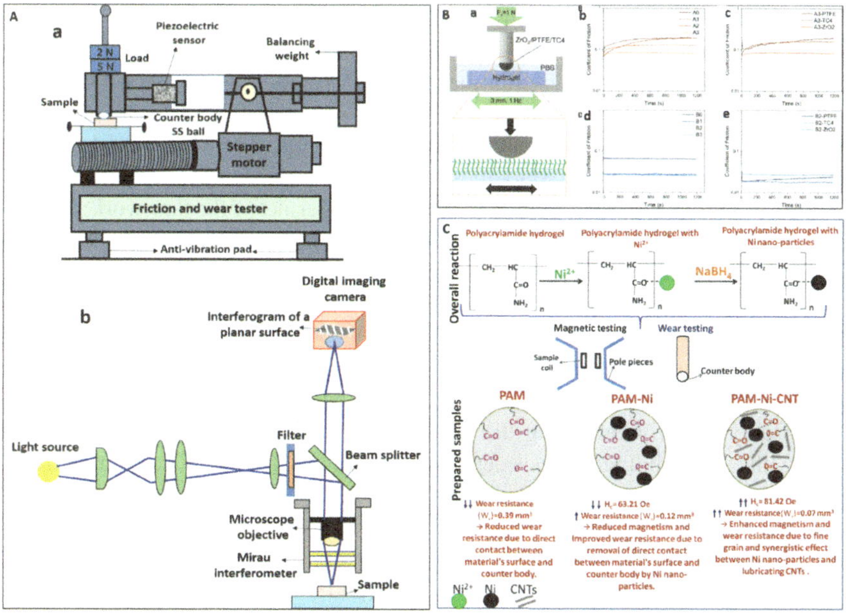

Fig. 4.6 **A** Diagram showing **a** fretting wear setup and **b** optical profilometry operation. Reproduced with permission from Ref. [68]. Copyright 2024 American Chemical Society (Open access under the license CC-BY-NC-ND). **B** KGM-modified hydrogels' tribological characteristics and possible uses. **a** KGM's lubricating system and friction test apparatus. ZrO_2 balls are used to measure the friction of **b** PVA-KGM and **d** PAM-KGM composite hydrogels. Friction test using PTFE, TC4, and ZrO_2 balls for the **c** A3 and **e** B2 groups. Reproduced with permission from Ref. [131]. Copyright 2025 Elsevier. **C** Schematic diagram to show the overall research work related to ferrogels. Reproduced with permission from Ref. [9]. Copyright 2021 Elsevier

Liquid crystals (LCs) have attracted a lot of interest due to their unusual blend of fluid-like behavior, which is typical of liquids, and long-range orientational order, which is characteristic of solids [110–112]. The combination of hydrogels and LCs presents intriguing opportunities for creating cutting-edge water-based lubricants. The polymer network of hydrogels restricts solutes in the aqueous phase, enabling convective diffusion and hydrodynamic flow participation during friction [113]. Numerous obstacles stand in the way of current research employing hydrogels as a static matrix to encapsulate surfactant-stabilized LCs [114, 115]. The absence of an enriched LC layer with homeotropic alignment at the contact interface and competing surface anchoring by polar molecules limits the effectiveness of LC-based lubrication in tribological systems [116]. The tribological benefits of LCs cannot be completely exploited in the absence of a solid lubricating coating or strong intermolecular contacts [117]. To bridge this gap, Gong et al. [118] investigated the possibility of γ-cyclodextrin (γ-CD) as a water-based lubricant by dynamically stabilizing 4′-pentyl-4-cyanobiphenyl (5CB) molecules inside poly(vinyl alcohol) hydrogels. The channel-type crystalline structure of γ-CD and the ordered molecular arrangement of 5CB were both conserved in the resulting 5CB/γ-CD inclusion complexes. Incorporating the 5CB/γ-CD inclusion complexes into the hydrogel network reduced hydrogel swelling and raised the compressive modulus because they interact hydrophobically with the polymer backbone. Tribological tests revealed velocity-dependent friction behavior. At low sliding velocities (1.3 cm/s), partially exposed 5CB/γ-CD inclusion complexes increased interfacial friction via hydrophobic interactions. Shear-induced alignment of 5CB promoted fluid flow and interfacial slippage at higher velocities (2.6 cm/s), lowering the coefficient of friction. By stabilizing liquid crystals inside a hydrogel network, this work presents an alternative method for developing water-compatible, versatile tribological materials for dynamic friction control applications. Dual cross-linked zinc-coordinated copolymer hydrogels are demonstrated to be easily synthesized in a single pot by Nageshwar and the research team [119]. The copolymer hydrogels' network structure is purposefully created by Zn^{2+} ions' dynamic and reversible physical cross-linking and concurrent covalent cross-linking using a covalent cross-linker, methylene bisacrylamide. Comparing the HGel-$Zn(NO_3)_2$ sample to the original HGel sample, the shear sliding analysis demonstrates the HGel-$Zn(NO_3)_2$ sample's dominating sticky behavior. With its advantageous adhesive nature, strong mechanical capabilities, effective self-recovery, and sufficient water content, this easy dual cross-linked hydrogel, HGel-$Zn(NO_3)_2$, appears to be a very attractive candidate to simulate articular cartilage. Hydrogels are very malleable and sticky, just like genuine cartilage and tissues. The composition, specimen thickness, substrate type, substrate roughness, normal load, ambient temperature, and humidity all often influence the hydrogels' adhesive properties [120, 121]. Sticky hydrogels are naturally sticky and have characteristics similar to cartilage tissue. Only after the hydrogel is fully integrated with the implant can it serve as a substitute for articular cartilage [122]. Hydrogel should have a sticky quality to solve this. In the event that cartilage is injured, adhesive hydrogels can provide mechanical joint support. Adhesive hydrogels provide durable tissue-implant integration, improving therapeutic results [123].

D-mannose and D-glucose are joined in β-1,4 glycosidic linkages to form Konjac glucomannan (KGM), a plant polysaccharide. It has not been documented to regulate hydrogel mechanical and lubricating performance, but it has application potential in wound healing and medication delivery [124–126]. It has been demonstrated that the polysaccharides decrease the CoF by forming a hydration layer [127, 128] and increase the compression resistance of hydrogels by hydrogen bonding [129, 130]. KGM can control the performance of hydrogels because of its strong hydrogen bonding and hydration properties. However, the system stability of various crosslinked hydrogels after KGM was introduced is yet unknown; therefore, the assumptions need to be empirically validated. Additionally, a major obstacle during the preparation stage is the high viscosity brought on by KGM. For this purpose, Shao et al. [131] proposed the addition of varying KGM concentrations in the PVA and PAM hydrogels and examined their tribological characteristics. As the KGM concentration increases, the CoF for PVA-KGM lower samples in Fig. 4.6B b, c are 0.175, 0.158, 0.124, and 0.083. PAM hydrogel has a lower initial CoF for PAM-KGM lower samples, as shown in Fig. 4.6B d, e, and KGM effectively contributes to lubrication, going from 0.066 to 0.027, 0.026, and 0.027. The performance of the macromolecular hydrophilic layer is correlated with its thickness, as seen in Fig. 4.6B a, where a portion of the KGM that is still present on the hydrogel surface forms a hydration layer by hydrogen bonding with water molecules. Both composite hydrogels' lubricating qualities are successfully enhanced by KGM, which also lowers the minimum CoF of both kinds of samples to less than 0.10. In a recent work, Vukonic et al. [132] revealed the unique functions of these biopolymers in modifying the characteristics of double-network (DN) hydrogels based on PVA, offering important new information on their benefits and limitations for tissue engineering uses. Agarose allows macroporous formations and increases compressive strength, making it perfect for load-bearing applications. Silk fibroin (SF) has remarkable water retention and energy dissipation qualities, while chitosan offers great tribological performance and elastic recovery. Agarose and chitosan (CS) work in concert to create a multilayer (ML) hydrogel architecture that integrates structural reinforcement with increased flexibility. A ball-on-disc tribometer was used to conduct wear testing of the materials. Effective use as artificial cartilage requires a low coefficient of friction (COF) to guarantee durability and reduce wear in vivo. Important information on the frictional behavior of PVA-based hydrogels under physiologically simulated settings may be obtained by tribological assessment. Interestingly, with an average COF of 0.09, the basic PVA hydrogel outperformed the others. Among the additions, the average COF for SF and CS was somewhat higher at 0.11, and it increased even more to 0.12 with the addition of Agar. These low COF values are caused by the hydrogels' high water content and lubricating qualities, which replicate the behavior of biological tissues in the natural world. These features highlight the potential of hydrogels for applications requiring reduced wear and friction while addressing a major limitation of conventional implant materials.

Using a two-step in-situ swelling technique, a biomimetic bilayer hydrogel made of PAM (polyacrylamide) and PAM/SA (sodium alginate) was created, mimicking the distinct structure of articular cartilage [133]. The studies on the tribological performance of the PAM-PAM/SA bilayer hydrogel have shown that it operates with very little friction, a coefficient of friction as low as 0.032, even when subjected to high loads. Finally, by using the different swelling properties of the PAM-PAM/SA bilayer hydrogel's several layers, a self-bending biomimetic cartilage that can adjust to complex joint surfaces is produced. This conformal, mechanically strong, and highly lubricating biomimetic cartilage offers a practical solution for joint disorders and cartilage abnormalities.

Ti6Al4V alloys have found use in the field of hard tissue implantation. Their low wear resistance, however, has been identified as a significant contributing cause to the aseptic loosening of implants. Its use in the biological domains was therefore constrained. A chitosan-MXene hydrogel coating was created and applied to the surface of Ti6Al4V alloys in the work done by Li and team [134]. Tests were conducted on the hydrogel coating to examine tribological performance in simulated bodily fluid (SBF) solution. The findings demonstrated the outstanding tribological performance of the CS-MXene-0.3% hydrogel coating, as evidenced by the 78.33% decrease in friction coefficient. A possible surface modification technique for titanium alloy bone implants was presented in this work.

Research into ferrogel tribology, the study of wear, lubrication, and friction in surfaces that interact with them, has potential uses in a number of domains, including medication administration and joint replacement. Ferrogels' tribological characteristics are not as thoroughly studied as their magnetism and stimulus–response behavior. According to research, ferrogels can have intriguing tribological properties, especially when mixed with other substances like carbon nanotubes (CNTs). For instance, Awasthi et al. [9] reported ferrogel for magnetic biosensors and magnetic field-sensitive transformers. Nickel nanoparticle-embedded polyacrylamide hydrogel was enhanced with CNTs to achieve great wear resistance, strength, and magnetic sensitivity. Because of their lubricating properties and the synergistic impact, they generate with Ni nanoparticles, it is also shown that the wear rate decreased by around 40% when CNTs were added to the ferrogel. As a result, ferrogel with CNT added exhibits more magnetism and less wear damage, making it a potentially sophisticated polymer composite for use in drug administration and magnetic sensors. After the polyacrylamide and $NiCl_2$ interacted to create a hydrogel saturated in green Ni(II) ions, the ions were reduced by the reducing agent $NaBH_4$ to create black-looking Ni nanoparticles. Consequently, three types of PAM, PAM-Ni, and PAM-Ni-CNT samples were created. Since there was no direct contact between the material's surface and counter body, PAM-Ni demonstrated superior wear resistance ($Wv = 0.12\ mm^3$) compared to pure PAM ($Wv = 0.39\ mm^3$) in the evaluation of the manufactured samples' tribological and magnetic performances. However, the highest wear resistance for PAM-Ni-CNT ($Wv = 0.07\ mm^3$) was achieved by a protective layer produced by lubricating CNTs and a synergy made by finely contacted Ni and CNTs (Fig. 4.6C) in the PAM matrix.

The studies, as mentioned above, showcase recent tribological advances of hydrogels and ferrogels. However, some other research on friction and wear characteristics of HG-FGs is given in Table 4.2 (2020–2023) for a more detailed overview of readers.

Table 4.2 Current studies on the tribological characteristics of hydrogels for biomedical applications [135]

Materials	Synthesis technique	Compact strength (MPa)	Compact module (MPa)	COF	Results	References
PVA/GO-HA	3D printing technique	0.93	0.96	0.069	Improved dynamic viscosity	[136]
DT/Ti6Al4V-PSBMA	Crosslinking	–	2.35	0.08	Increased tribological performance due to hydration lubrication mechanisms	[137]
Agar/hydrogel	Crosslinking	62.35	–	0.303	Low friction	[138]
PEEK/hydrogel/polymer brush	Crosslinking	–	5.42	0.19	High friction resistance	[139]
Gelatin-based composite hydrogels	In-situ injectable hydrogel	–	–	0.027	Reduced COF due to the functionalized microspheres	[140]
HPX/PVA hydrogels	Crosslinking	0.9	0.325	0.029	Reduced wear due to hydrogen-bonding interaction	[141]
Gelatin-based composite hydrogels	Crosslinking	0.788	–	0.014	High wear resistance	[142]
PVA-25%PMEDAH (molecular weight)	Free radical polymerization	0.7	–	0.04	Enhanced tribological performance due to interfacial interaction	[143]
PEEK-PAA-PVA-METAC	Free radical polymerization	–	2.82	0.062	Lubrication mechanism for tribological behavior	[144]
p(HEMA-co-DMAA) lipid hydrogels	Crosslinking	2.8	2.0	0.026	Non-toxic groups like methyl and carbonyl	[145]
PEEK-PVA-tannic acid hydrogel	Crosslinking	–	7	0.06	The low-friction and self-healing properties	[146]

(continued)

Table 4.2 (continued)

Materials	Synthesis technique	Compact strength (MPa)	Compact module (MPa)	COF	Results	References
PVA/SA/NaCl	Simple freeze–thaw	–	–	0.03–0.1	Minimal toxicity and better lubricating qualities	[147]
Bilayer hydrogels	Magnetic field orientation, freeze–thaw and annealing process	5.21	4.06	0.032	Lubricating composites	[148]
AA-AM hydrogel layer on PEEK substrates	Dual cross-linked	–	–	0.061	Medical wear-resistant materials based on PEEK, especially for orthopedic implants	[149]
Top soft hydrogel/brush composite	Hybrid covalently cross-linked network	–	–	0.025	The proposed technology inspires the development of innovative bionic lubricating materials with a substantial load-bearing capacity	[150]
Soft (hydrogel)-hard (Ti6Al4V)	Freezing–thawing and annealing treatments	2.2	0.5	0.078	The Ti6Al4V hydrogel materials exhibit ultra-high interfacial toughness and remarkable tribological properties	[151]
Multilayer bionic composite	Freezing–thawing and annealing treatments	3.1	4.47	0.035	Because of its high strength and minimal friction, it is a wonderful joint material	[152]

(continued)

Table 4.2 (continued)

Materials	Synthesis technique	Compact strength (MPa)	Compact module (MPa)	COF	Results	References
Fiber/PVP-PHEMA	Chemical-crosslinking hydrogels mixed with micron-scale fibers	9	17	0.135	The composite hydrogels are more lubricating, load-bearing, and anti-wear than the traditional hydrogel system. They also hold great promise for the creation of biomimetic materials	[153]
PVA/PEG–glycerol hydrogel	Mixing physical crosslinking	22	–	0.18	Given its ease of production and favorable mechanical and lubricating properties, this hydrogel material may find use in the repair of articular cartilage lesions	[154]
Hairy and slippery polyoxazoline-based copolymers	Crosslinking	–	–	0.3	These findings suggest that, in the wake of degenerative diseases like osteoarthritis, cartilage can be preserved and its lubricating properties restored by carefully modifying the composition and structure of tissue-reactive copolymers	[155]
(AM-co-AA) sugarcane composite hydrogel	Hydrogen bonding interaction	–	–	0.048	Anisotropic sugarcane composite hydrogels have a lot of potential applications like flexible sensors, bioinspired prosthetic skin, and cartilage lubricants	[156]

(continued)

Table 4.2 (continued)

Materials	Synthesis technique	Compact strength (MPa)	Compact module (MPa)	COF	Results	References
MTi6Al4V-zwitterionic hydrogel	Crosslinking	–	0.712	0.13	The surface of the Ti6Al4V alloy may be functionalized for artificial joints in a useful manner with this novel MTi6Al4V-PVA-PAM-PMPC composite structure	[157]
PAAm/PVA hydrogel-Ti6Al4V	Crosslinking	0.45	0.113	0.05	The mechanical and tribological properties of a 1:3 PAAm:PVA hydrogel are perfectly balanced	[158]
PVA/borax/PEO hydrogel	One-pot method	–	–	0.17	This new technique will help us understand how to enhance the friction properties of hydrogels for usage in biological or engineering applications	[159]
Triple-network (PVA/SB TN) hydrogel	Crosslinking	0.92	–	0.091	More importantly, the PVA/0.4SB triple-network hydrogel's friction coefficient is similar to that of native cartilage under the identical test circumstances	[160]
PVA–pMPDSAH double-network hydrogel	Physical crosslink	–	–	0.11	This blended composite of cross-linked hydrogel provides a novel method for researching biomimetic articular cartilage	[161]

(continued)

Table 4.2 (continued)

Materials	Synthesis technique	Compact strength (MPa)	Compact module (MPa)	COF	Results	References
Organic solvents—PVA/PEG hydrogels	Crosslinking	–	–	0.118	Improved tribological properties of PVA gels for articular cartilage repair	[162]
PVA-BA/PEG hydrogel	Cast-dry method and freeze–thaw	1.35	2.3	0.04	The mechanical performance and aqueous lubrication of all PVA-BA/PEG hydrogels produced with the two methods were enhanced	[163]
Micropored pHEMA hydrogels	Photolithography	–	–	0.75	These results suggest that pHEMA hydrogels with carefully designed micropore textures might be useful in biological applications	[164]

4.1.4 Durability of HG-FGs

3.5 wt. % solution of NaCl, borate buffer solution with a combination of $Na_2B_4O_7 \cdot 10H_2O$, KNO_3, and H_3BO_3, or simulated body fluid (SBF, for biological samples) can be used to assess the materials' durability in harsh conditions, using a potentiostat. The samples, saturated calomel electrode, and platinum are used as the working electrode, reference electrode, and counter electrode, respectively. The corrosion test may be performed using a variety of testing techniques, including electrochemical impedance spectroscopy (EIS), linear polarization, potentiodynamic polarization, and open circuit potential (OCP). The potential of the working electrode in relation to the reference electrode at which an anodic and cathodic reaction rate equilibrium is reached in the absence of an applied voltage is known as the open circuit potential, or OCP. At this point, the electrochemical measurements begin.

The linear polarization resistance is given by the figure's slope. On the other hand, corrosion current (i_{corr}) is the term used to describe the anodic or cathodic current value at OCP. Potentiodynamic polarization (PD) may be scanned from OCP at a specific scan rate and potential range. The anodic and cathodic slopes (β_{a} and β_{c}, respectively) are described by the classic Tafel figure using Tafel extrapolation. The Stern Geary equation is used to calculate the polarization resistance from PD (R_{PD}) [165].

$$R_{\mathrm{PD}} = \frac{\beta_{\mathrm{a}}\beta_{\mathrm{c}}}{2.303\, i_{\mathrm{corr}}(\beta_{\mathrm{a}} + \beta_{\mathrm{c}})}, \tag{4.10}$$

where i_{corr} represents the corrosion current density and Tafel slopes using β_{a} (anodic), β_{c} (cathodic). The sample's corrosion rate (C_{R}) was estimated using Faraday's equation.

$$C_{\mathrm{R}} = K\left(\frac{i_{\mathrm{corr}}}{\rho}\right)E_{\mathrm{W}}, \tag{4.11}$$

where, K (3.27×10^{-3} mm g/μA cm year) is a constant while ρ shows density, and E_{W} is the equivalent weight of the samples. Additionally, corrosion rates are denoted by w_1 (no reinforcements) and w_2 (presence of reinforcements). This may be used to calculate how reinforcements affect the percentage corrosion inhibition efficiency ($\%_{\mathrm{CIE}}$) [166].

$$\%_{\mathrm{CIE}} = \left[1 - \frac{w_2}{w_1}\right] \times 100 \tag{4.12}$$

Apart from potentiodynamic polarization (PD) and linear polarization (LP), electrochemical impedance spectroscopy is employed to evaluate the components of the Nyquist plot, Bode plot, and equivalent circuits in order to further quantify corrosion resistance.

Existing studies on hydrogels and ferrogels focus mainly on nanoparticle leaching, oxidative stability of iron particles, and pH-dependent degradation. Only a few have adapted electrochemical methods, such as modified electrochemical impedance spectroscopy (EIS), to HG-FGs. Overall, while hydrogel degradation is well-documented, true corrosion studies, particularly electrochemical corrosion, remain a relatively underexplored but increasingly important field. In this context, the research work was done by considering corrosion studies of gels in simulated body fluid [167]. PAM-based robust and resilient dimer (PAM-Ag and PAM-BNOH) and trimer (PAM-Ag-BNOH) nanocomposites were synthesized by encapsulating silver (Ag) and hydroxylated hexagonal boron nitride (BNOH), inspired by cartilage tissue. Hydrogels were electrochemically tested in a simulated bodily fluid to determine their long-term durability. Trimer hydrogels demonstrated a greater corrosion resistance ($i_{corr} = 2.65 \times 10^{-5}$ A cm^{-2}). Thus, the robust hydrogel composite that was created has the potential to be a leading biomaterial for the regeneration of soft tissue, such as articular cartilage. The current work paves the way for the development of intelligent self-healing nanocomposites, which are widely employed in cartilage tissue engineering.

4.1.5 Magnetism of Ferrogels

The samples' magnetic coercivity (H_c), remanence (M_r), and saturation magnetization (M_s) are measured using a vibrating sample magnetometer (VSM). The fundamental principle of VSM is Faraday's law of magnetic induction, which says that an electromagnetic force is produced in a coil when a change in flux occurs through the coil. Based on atomic force microscopy, the magnetic force microscopy method is used to determine the magnetization hysteresis curves of the ferromagnetic particles in the samples. There is surface magnetism in these domains. The domain borders are represented by black regions, while the primary domains are represented by bright patches. We define the width of the domain wall as the average of the MFM amplitude maxima. It is also possible to get the magnetostatic energy per unit area (e_{ms}) using the domain wall width (W) and magnetic saturation (M_s).

$$e_{ms} = 0.85\, W (M_s)^2. \tag{4.13}$$

Behavioral and electrophysiological impairment are the outcomes of spinal cord injury (SCI). Although electrical stimulation (ES) is thought to be a successful therapy for moderate SCI, tissue abnormalities that impair electrical conduction make ES ineffective for severe SCI. Therefore, a possible treatment approach is to fill in the deficiencies and restore electrical conductivity in the spinal cord using conductive materials. In a recent work, by mixing conductive reduced graphene oxide (rGO) and magnetic Fe_3O_4 nanoparticles using ultrasonic technology and encasing them in gelatin methacryloyl (GelMA) and decellularized extracellular matrix (dECM), Yan et al. [168] produced a conductive composite hydrogel, rGO/Fe_3O_4/dECM@GelMA. By using a magnetic field to position themselves within the hydrogel, the rGO/Fe_3O_4

complexes gave the hydrogel an oriented electrical conduction function. The goal of this research is to enhance the effectiveness of ES treatment in cases of severe spinal cord injuries and facilitate the recovery of spinal cord function by combining it with a conductive tissue-engineered scaffold.

During the process of polymerization and cross-linking, a novel ferrogel with chemically controlled magnetism was produced by dispersing magnetic nanoparticles (MNPs) in a hydrogel composed of a pH-responsive sulfamethazine-based polymer [169]. In the rheological matrix, the MNPs exhibit dynamic rotations or oscillations when subjected to alternating current (AC) magnetic fields. The way the integrated MNPs behave in terms of AC magnetization is controlled by the hydrogel's swelling in response to pH changes between 7 and 12. Higher swelling ratio ferrogels exhibit better Brownian relaxation during AC magnetization. A sensitivity of 5.3×10^{-6} mol/L or less has been achieved by using ferrogel as a sensing label and analyzing the pH response of the AC magnetization to detect ammonia gas.

4.1.6 Biological Analysis of HG-FGs

Biological characterization is the process of creating, maintaining, and effectively controlling certain biological traits within a given system. The scaffold matrix should assist vascularization while allowing sufficient space for tissue formation and offer an environment that supports cellular activity, including adhesion, migration, and functioning. Promoting processes that are similar to the organism's physiology is the main objective of the cell carrier. One of the most important aspects of scaffold engineering is managing the dynamic balance between bone tissue development and breakdown. While early breakdown can promote the creation of physiologically more powerful structures like connective tissue, prolonged resorption can impede bone regrowth. Therefore, the ideal substance should be in harmony with the biological processes of the tissue in which it is used [170].

Because of its many biological characteristics, such as its immunomodulatory, anti-inflammatory, and antioxidant qualities, yeast β-glucan shows great promise as a wound healing regent. However, the limited water solubility of yeast β-glucan restricts its use. A variety of carboxymethylated yeast β-glucans (CMGs) with varying degrees of substitution were synthesized for the investigation done by Xu and co-workers [171]. Using a zebrafish model, the impact of substitution degree on the biological activities of carboxymethylated yeast β-glucan was examined. In zebrafish, CMG3 with replacement degree 0.55 exhibited immunomodulatory, antioxidant, and anti-inflammatory properties. Additionally, CMG3 demonstrated the capability to stimulate caudal fin regeneration and angiogenesis. L929 cells were used to assess the cytotoxicity of CMG3, and a scratch test was used to examine the impact of CMG3 on cell migration. The in vitro antioxidant activity and hemolysis ratio of CMG3 were investigated. Then, using a straightforward technique, CMG3 was utilized to create a sprayable hydrogel, and a model of diabetic mice produced by streptozotocin was employed to examine the wound-healing capacity. According to the findings, a sprayable hydrogel based on CMG3 might hasten wound healing in diabetic mice and affect the expression of biomarkers linked to angiogenesis, inflammation, and macrophage polarization. To investigate the wound healing of CMG3-based sprayable hydrogel, a full-thickness skin defect mouse model was used. Streptozotocin was used to cause type I diabetes in mice. Diabetic mice's wounds were sprayed with varying concentrations of CMG3-based sprayable hydrogel (2.5, 5.0, and 10 mg/mL), and at days 0, 3, 6, 12, and 18, the healing process was captured on camera. Due to the inflammatory response of all groups, there was no discernible change in the wounds on day three, as seen in Fig. 4.7A, B. However, following 6, 12, and 18 days of hydrogel therapy, the wound healing process was considerably accelerated. When compared to the control group (61.79 $\pm$ 1.83%), the wound treated with 10 mg/mL hydrogel had a reduced remaining area at day 12 (Fig. 4.7C). The findings of the studies at day 18 were comparable to those at day 12, suggesting that the CMG3-based sprayable hydrogel may have the capacity to repair wounds. The hydrogel concentration has a major impact on how quickly wounds heal. A minor amount of wound was left behind due to the high hydrogel content. In comparison to wounds treated with 2.5 and 5.0 mg/mL hydrogel (6.14 $\pm$ 1.03% and 4.17 $\pm$ 0.49%, respectively), the wound treated with 10 mg/mL CMG3-based sprayable hydrogel had a reduced remaining area (1.74 $\pm$ 0.54%) at day 18.

In the clinic, treating full-thickness wounds infected with germs is still quite difficult. In a recent study, marine snail peptide (Tyr-Ile-Ala-Glu-Asp-Ala-Glu-Arg), dialdehyde carboxymethyl cellulose, and carboxymethyl chitosan were used as starting ingredients to create a HYP hydrogel [172]. Reactive oxygen species in wound sites may be eliminated by the marine snail peptide with strong antioxidant action, reducing the overabundance of inflammation. HYP has strong injectable and self-healing properties made possible by the dynamic Schiff-base linkages. Wound healing was aided by HYP's unique porosity structure, superior rheological characteristics, and appropriate gelation time. According to in vitro biological investigations, HYP demonstrated enhanced antioxidant and antibacterial properties, a low hemolysis ratio, and high biocompatibility. In a bacterially contaminated full-thickness skin defect rat model, an in vivo investigation showed that HYP might facilitate wound healing. There were less inflammatory cells, new hair follicles, and clear collagen deposition in the wound tissues. Significant improvements were observed in the expression of biomarkers linked to inflammation and angiogenesis, including IL-6, IL-10, CD31, and α-SMA. As a result, HYP hydrogel shown excellent potential for use as a dressing for bacterially infected wounds.

Table 4.3 presents an extended collection of significant biological studies conducted between 2020 and 2025, focusing on hydrogel and ferrogel systems in biomedical applications. These studies offer insights into recent innovations in areas such as tissue engineering, drug delivery, neural repair, and wearable technology. It is important to note that these studies are in addition to those previously summarized in the main text, providing further depth and diversity to the field's latest advancements.

Certain parts of this chapter have been adapted from the open-access article [68] under the Creative Commons CC BY-NC-ND 4.0 license. The content has been paraphrased and appropriately cited in accordance with the license terms.

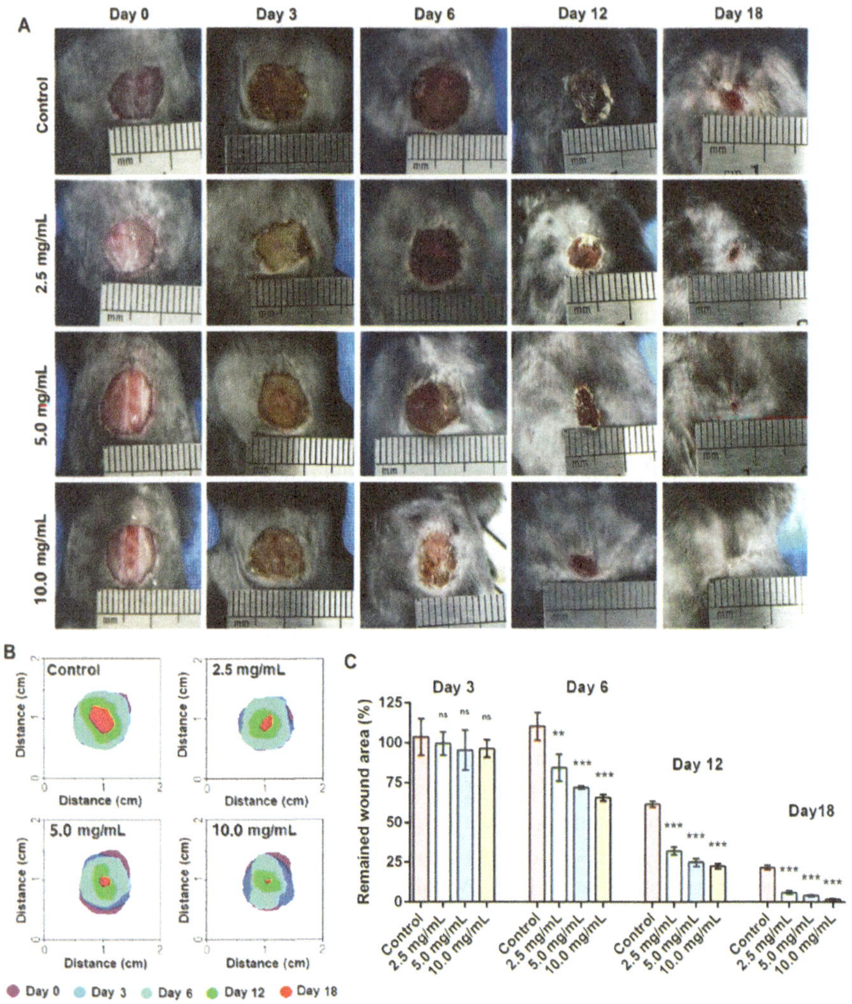

Fig. 4.7 **A** Pictures of diabetic mice's wounds after they were treated with varying amounts of sprayable hydrogel based on CMG3 (2.5, 5.0, and 10 mg/mL). **B** Image-Pro Plus 5.1 software was used to evaluate wound images taken on days 0, 3, 6, 12, and 18. **C** Wound regions that remained after days 3, 6, 12, and 18. On day zero, the wound regions were designated as 100%. Reproduced with permission from Ref. [171]. Copyright 2025 Elsevier

4.2 Summary and Outlook

This chapter explores recent advances in the development and multiphysical characterization techniques of hydrogels and ferrogels, focusing on their mechanical, tribological, electrochemical, magnetic, and biological properties. These soft, water-rich materials, especially when embedded with magnetic nanoparticles in the case of

Table 4.3 Recent biological studies on hydrogels and ferrogels (2020–2025)

Year	Theme of study	Hydrogel systems	Application	Key findings	References
2020	3D printing of self-healing ferrogel	Glycol chitosan, oxidized hyaluronate, SPIONs	Tissue engineering, 3D printing	Developed self-healing ferrogels with magnetic responsiveness suitable for 3D printing and drug delivery	[173]
2020	Magnetic sensitive hydrogels in tissue engineering	Magnetic hydrogels with iron oxide nanoparticles	Tissue engineering	Magnetically responsive hydrogels offer biocompatibility and remote control for regenerative medicine	[174]
2021	Injectable nanocomposite hydrogels for cardiac repair	Nanocomposite hydrogels	Cardiac tissue engineering	Reduced ROS levels and improved cardiac function after myocardial infarction	[175]
2021	3D-printed alginate-based hollow fiber scaffold	Alginate-based magnetic hydrogels	Drug/cell delivery	Enabled magnetically driven, on-demand drug and cell release	[176]
2022	Piezoresistive MXene/silk fibroin hydrogel	MXene, silk fibroin	Bone regeneration	Re-established electrical microenvironment to enhance bone repair	[177]
2022	Injectable macroporous ferrogel beads	Macroporous ferrogels	Magnetically actuated drug delivery	Achieved structural stability and efficient magnetic drug release	[178]
2023	Piezoelectric hydrogel for osteoarthritis	Poly-L-lactic acid fibers, collagen	Osteoarthritis therapy	Promoted cartilage repair using ultrasound-activated piezoelectric hydrogel	[179]
2023	Curcumin-loaded GelMA hydrogel	Gelatin methacryloyl, curcumin	Cartilage regeneration	Anti-inflammatory hydrogel improved cartilage tissue regeneration	[180]
2024	Ecofriendly magnetic gel beads	Carboxymethylcellulose, iron oxides	Environmental remediation	Removed diclofenac from water, showcasing sustainability and reuse	[50]
2024	pH-sensitive magnetic hydrogel	PVA, starch, $MnFe_2O_4$ NPs	Drug delivery (5-FU)	Enabled sustained 5-FU release with high encapsulation efficiency	[181]
2024	Injectable smart responsive hydrogels	Alginate, SPIONs, gelatin	Drug delivery, hyperthermia	Magnetic field triggers localized drug release and tumor treatment	[182]

(continued)

Table 4.3 (continued)

Year	Theme of study	Hydrogel systems	Application	Key findings	References
2024	Biomimetic hydrogel with amyloid fibers	HET-s amyloid fibers	Neural tissue engineering	Supported long-term culture of neurons; potential for brain repair	[183]
2025	3D-printed organohydrogel sensors	Methacrylic anhydride-κ-carrageenan	Wearable biosensors	Created biocompatible, self-adhesive hydrogel sensors for health monitoring	[184]
2025	Photothermal nanohybrid hydrogel	Polydopamine NPs, methacryloyl gelatin	Bone repair, cancer therapy	Photothermal effect used for dual-purpose tissue regeneration and cancer targeting	[185]
2025	Stimuli-responsive hydrogels for OA therapy	Multi-functional smart hydrogels	Osteoarthritis	Surveyed innovations in smart hydrogels for OA, showing enhanced therapeutic outcomes	[186]

ferrogels, have emerged as key players in fields ranging from soft robotics and drug delivery to tissue engineering and bioelectronics.

1. Mechanical Studies

Recent studies have emphasized improving the mechanical robustness of hydrogels and ferrogels without compromising their flexibility or responsiveness. Techniques such as double-network structuring, nanofiller reinforcement, and interpenetrating polymer networks have led to significant improvements in tensile strength, toughness, and resilience. Ferrogels, in particular, exhibit tunable stiffness under magnetic fields, enabling active mechanical modulation in situ.

2. Tribological Studies

Hydrogels are naturally low-friction materials, making them ideal for artificial cartilage and lubricating coatings. Tribological studies have focused on optimizing surface interactions through hydration layers, polymer chain mobility, and nanoparticle incorporation. Ferrogels offer enhanced control of friction and wear through magnetic field modulation, enabling switchable lubrication and adaptive interfaces.

3. Electrochemical Studies

Electrochemical studies have expanded the potential of hydrogels as flexible electrodes and biosensors. Incorporation of conductive polymers (e.g., polyaniline, PEDOT:PSS) and ionic liquids into hydrogel matrices has led to improvements in ionic conductivity, charge storage, and redox stability. Ferrogels further introduce dynamic control of ion transport and electroactivity under magnetic stimuli, supporting applications in soft energy devices and actuators.

4. Magnetic Studies

Ferrogels, embedded with iron oxide or similar nanoparticles, exhibit responsive behavior under external magnetic fields. Recent work has demonstrated spatial and temporal control over deformation, heat generation (magnetothermal), and drug release. The magnetic characteristics depend on nanoparticle dispersion, polymer crosslinking density, and external field parameters, making them tunable for targeted biomedical or robotic applications.

5. Biological Studies

Hydrogels are widely recognized for their excellent biocompatibility and ability to mimic natural extracellular matrices. Recent research has improved their biofunctionality by incorporating growth factors, antibacterial agents, and cell-adhesion peptides. Ferrogels have shown promise in targeted drug delivery and remote-controlled cell stimulation due to their magneto-responsive behavior. Studies also focus on minimizing cytotoxicity from magnetic particles while enhancing regenerative capacity and immunomodulation.

Thus, hydrogels and ferrogels represent a dynamic class of materials with highly tunable properties that span multiple functional domains. Advances in synthesis and characterization have enabled significant progress in their design of adaptive, biointegrated, and intelligent systems. The synergy between mechanical integrity, electrochemical functionality, and magnetic responsiveness continues to drive innovation in soft matter engineering.

References

1. T.L. Burnett, P.J. Withers, Completing the picture through correlative characterization. Nat. Mater. **18**, 1041–1049 (2019). https://doi.org/10.1038/s41563-019-0402-8
2. S. Mourdikoudis, R.M. Pallares, N.T.K. Thanh, Characterization techniques for nanoparticles: comparison and complementarity upon studying nanoparticle properties. Nanoscale **10**, 12871–12934 (2018). https://doi.org/10.1039/C8NR02278J
3. S. Ebnesajjad, Surface and material characterization techniques, in *Handbook of Adhesives and Surface Preparation* (Elsevier, 2011), pp. 31–48. https://doi.org/10.1016/B978-1-4377-4461-3.10004-5
4. T. Funami, M. Hiroe, S. Noda, I. Asai, S. Ikeda, K. Nishinari, Influence of molecular structure imaged with atomic force microscopy on the rheological behavior of carrageenan aqueous systems in the presence or absence of cations. Food Hydrocoll. **21**, 617–629 (2007)
5. Y. Feng, W.H. Roos, *Atomic Force Microscopy: An Introduction* (2024), pp. 295–316. https://doi.org/10.1007/978-1-0716-3377-9_14
6. H.-B. Yang, Y.-X. Lu, X. Yue, Z.-X. Liu, W.-B. Sun, W.-P. Zheng, Q.-F. Guan, S.-H. Yu, Multiscale integral synchronous assembly of cuttlebone-inspired structural materials by predesigned hydrogels. Nat. Commun. **16**, 62 (2025). https://doi.org/10.1038/s41467-024-55344-1
7. S. Awasthi, Hierarchical heterostructures of metal alloy nanocoatings: expanding nanoplatforms for versatile synthesis and diversified applications. Coord. Chem. Rev. **527**, 216397 (2025). https://doi.org/10.1016/j.ccr.2024.216397
8. S. Awasthi, S.K. Pandey, J.K. Gaur, C. Srivastava, Load-bearing study and interfacial interactions of hydroxyapatite composite coatings for bone tissue engineering. Mater. Chem. Front. **6**, 3731–3747 (2022). https://doi.org/10.1039/d2qm00898j
9. S. Awasthi, J.K. Gaur, M.S. Bobji, Advanced ferrogels with high magnetic response and wear resistance using carbon nanotubes. J. Alloys Compd. **848**, 156259–156269 (2020)
10. A. Himawan, A. Korelidou, A.M. Pérez-Moreno, J.L. Paris, J. Dominguez-Robles, L.K. Vora, A.D. Permana, E. Larrañeta, R. Graham, C.J. Scott, R.F. Donnelly, Formulation and evaluation of PVA-based composite hydrogels: physicochemical, leachables, and in vitro immunogenicity studies. J. Mater. Chem. B **13**, 2431–2445 (2025). https://doi.org/10.1039/D4TB02181A
11. C.W. Bunn, Crystal structure of polyvinyl alcohol. Nature **161**, 929–930 (1948). https://doi.org/10.1038/161929a0
12. B.G. Colvin, Crystal structure of polyvinyl alcohol. Nature **248**, 756–759 (1974). https://doi.org/10.1038/248756a0
13. A. Himawan, Q.K. Anjani, U. Detamornrat, L.K. Vora, A.D. Permana, R. Ghanma, Y. Naser, D. Rahmawanty, C.J. Scott, R.F. Donnelly, Multifunctional low temperature-cured PVA/PVP/citric acid-based hydrogel forming microarray patches: physicochemical characteristics and hydrophilic drug interaction. Eur. Polym. J. **186**, 111836 (2023). https://doi.org/10.1016/j.eurpolymj.2023.111836
14. R. Ramadan, A.M. Ismail, Structural and physical comparison between CS/PVP blend and CS/PVP/Sr-hexaferrite nanocomposite films. J. Inorg. Organomet. Polym. Mater. **33**, 2506–2516 (2023). https://doi.org/10.1007/s10904-023-02684-y

15. G. Hao, Y. Hu, L. Shi, J. Chen, A. Cui, W. Weng, K. Osako, Physicochemical characteristics of chitosan from swimming crab (*Portunus trituberculatus*) shells prepared by subcritical water pretreatment. Sci. Rep. **11**, 1646 (2021). https://doi.org/10.1038/s41598-021-81318-0
16. Y. Zhang, C. Xue, Y. Xue, R. Gao, X. Zhang, Determination of the degree of deacetylation of chitin and chitosan by X-ray powder diffraction. Carbohydr. Res. **340**, 1914–1917 (2005). https://doi.org/10.1016/j.carres.2005.05.005
17. F. Colmenero, Mechanical properties of anhydrous oxalic acid and oxalic acid dihydrate. Phys. Chem. Chem. Phys. **21**, 2673–2690 (2019). https://doi.org/10.1039/C8CP07188H
18. A. Xiang, C. Lv, H. Zhou, Changes in crystallization behaviors of poly(vinyl alcohol) induced by water content. J. Vinyl Addit. Technol. **26**, 613–622 (2020). https://doi.org/10.1002/vnl.21775
19. M.P. Illa, C.S. Sharma, M. Khandelwal, Tuning the physiochemical properties of bacterial cellulose: effect of drying conditions. J. Mater. Sci. **54**, 12024–12035 (2019). https://doi.org/10.1007/s10853-019-03737-9
20. J.-S. Park, H.-A. Kim, J.-B. Choi, H.-J. Gwon, Y.-M. Shin, Y.-M. Lim, M.S. Khil, Y.-C. Nho, Effects of annealing and the addition of PEG on the PVA based hydrogel by gamma ray. Radiat. Phys. Chem. **81**, 857–860 (2012). https://doi.org/10.1016/j.radphyschem.2012.02.005
21. N.A. Peppas, E.W. Merrill, Poly(vinyl alcohol) hydrogels: reinforcement of radiation-crosslinked networks by crystallization. J. Polym. Sci. Polym. Chem. Ed. **14**, 441–457 (1976). https://doi.org/10.1002/pol.1976.170140215
22. O.A. Zakurdaeva, S.V. Nesterov, N.A. Shmakova, G.K. Semenova, E.O. Sozontova, V.I. Feldman, Radiation-chemical synthesis of poly(vinyl alcohol) hydrogel containing dicyclohexano-18-crown-6. Nucl. Instrum. Methods Phys. Res. B **265**, 356–361 (2007). https://doi.org/10.1016/j.nimb.2007.09.003
23. V. Štejfa, V. Pokorný, J. Rohlíček, M. Fulem, K. Růžička, Polymorphism of anhydrous oxalic acid unravelled. J. Chem. Thermodyn. **160**, 106488 (2021). https://doi.org/10.1016/j.jct.2021.106488
24. A.K. Sonker, K. Rathore, R.K. Nagarale, V. Verma, Crosslinking of polyvinyl alcohol (PVA) and effect of crosslinker shape (aliphatic and aromatic) thereof. J. Polym. Environ. **26**, 1782–1794 (2018). https://doi.org/10.1007/s10924-017-1077-3
25. S. Basak, G. Packirisamy, Impact of magneto-mechanical actuation on cell differentiation: a study using wireless, 3D-printed device and a porous ferrogel. Small (2025). https://doi.org/10.1002/smll.202412112
26. Y. Li, Y. Li, J. Cao, P. Luo, J. Liu, L. Ma, G.-L. Gao, Z. Jiang, 4D-printed magnetic responsive bilayer hydrogel. Nanomaterials **15**, 134 (2025). https://doi.org/10.3390/nano15020134
27. H. Wu, G. Wu, L. Wang, Peculiar porous α-Fe_2O_3, γ-Fe_2O_3 and Fe_3O_4 nanospheres: facile synthesis and electromagnetic properties. Powder Technol. **269**, 443–451 (2015). https://doi.org/10.1016/j.powtec.2014.09.045
28. I. Boulhaia, M. Bañobre-López, N. Moulai-Mostefa, Formulation and characterization of ferrogels based on agar gum and octenyl succinic anhydride (OSA) starch. Starch Stärke **77** (2025). https://doi.org/10.1002/star.202300193
29. D. Kowalczuk, M. Pitucha, Application of FTIR method for the assessment of immobilization of active substances in the matrix of biomedical materials. Materials **12**, 2972 (2019). https://doi.org/10.3390/ma12182972
30. M. Raheem, M.R. Roknabadi, A. Kompany, Synthesis and characterization of diopside nanoparticles for drug delivery by sol-gel. Mustansiriyah J. Pure Appl. Sci. **3**, 135–143 (2025). https://doi.org/10.47831/mjpas.v3i1.256
31. M.P. Prieto, M. Pérez-Estébanez, S.S. Gómez, J.M. de la Roja, C.A. Pino, Evaluation of mucilages efficacy in the removal of a natural adhesive (gacha) from a canvas by ATR–FTIR. First results. J. Cult. Herit. **71**, 153–164 (2025). https://doi.org/10.1016/j.culher.2024.11.016
32. A. Sharma, G. Kroumpouzos, M. Kassir, H. Galadari, A. Goren, S. Grabbe, M. Goldust, Rosacea management: a comprehensive review. J. Cosmet. Dermatol. **21**, 1895–1904 (2022). https://doi.org/10.1111/jocd.14816

33. M.C. Marchitto, A.L. Chien, Mast cell stabilizers in the treatment of rosacea: a review of existing and emerging therapies. Dermatol. Ther. (Heidelb.) **11**, 1541–1549 (2021). https://doi.org/10.1007/s13555-021-00597-7
34. E.J. Wladis, A.P. Adam, Immune signaling in rosacea. Ocul. Surf. **22**, 224–229 (2021). https://doi.org/10.1016/j.jtos.2021.08.017
35. S.-H. Yoon, I. Hwang, E. Lee, H.-J. Cho, J.H. Ryu, T.-G. Kim, J.-W. Yu, Antimicrobial peptide LL-37 drives rosacea-like skin inflammation in an NLRP3-dependent manner. J. Investig. Dermatol. **141**, 2885-2894.e5 (2021). https://doi.org/10.1016/j.jid.2021.02.745
36. H. Chen, Z. Zhang, J. Qi, C. Cao, M. Lin, L. Lyu, D. Xu, Novel thermosensitive hydrogel encapsulated carvedilol for the treatment of rosacea. ACS Omega **10**, 7964–7972 (2025). https://doi.org/10.1021/acsomega.4c08884
37. Z. Bujňáková, E. Dutková, M. Baláž, E. Turianicová, P. Baláž, Stability studies of As_4S_4 nanosuspension prepared by wet milling in poloxamer 407. Int. J. Pharm. **478**, 187–192 (2015). https://doi.org/10.1016/j.ijpharm.2014.11.043
38. Y. Luo, Z. Teng, Y. Li, Q. Wang, Solid lipid nanoparticles for oral drug delivery: chitosan coating improves stability, controlled delivery, mucoadhesion and cellular uptake. Carbohydr. Polym. **122**, 221–229 (2015). https://doi.org/10.1016/j.carbpol.2014.12.084
39. N. Pešić, A. Dapčević, B. Ivković, K. Kachrimanis, M. Mitrić, S. Ibrić, D. Medarević, Potential application of low molecular weight excipients for amorphization and dissolution enhancement of carvedilol. Int. J. Pharm. **608**, 121033 (2021). https://doi.org/10.1016/j.ijpharm.2021.121033
40. J. Wei, C. Liu, L. Shi, Y. Liu, H. Lu, Highly stretchable, self-healing, adhesive, 3D-printable and antibacterial double-network hydrogels for multifunctional wearable sensors. Int. J. Biol. Macromol. **292**, 138813 (2025). https://doi.org/10.1016/j.ijbiomac.2024.138813
41. S. Zeng, M. Shen, L. Yang, Y. Xue, F. Lu, S. Chen, Self-assembled montmorillonite–carbon nanotube for epoxy composites with superior mechanical and thermal properties. Compos. Sci. Technol. **162**, 131–139 (2018). https://doi.org/10.1016/j.compscitech.2018.04.035
42. J. Zhang, L. Wang, A. Wang, Preparation and properties of chitosan-g-poly(acrylic acid)/montmorillonite superabsorbent nanocomposite via in situ intercalative polymerization. Ind. Eng. Chem. Res. **46**, 2497–2502 (2007). https://doi.org/10.1021/ie061385i
43. G.R. de Carvalho, A.M. Kudaka, R.A. Netto, C. Delarmelina, M.C.T. Duarte, L.M.F. Lona, Antiviral and antibacterial activity of sodium alginate/poly(diallyldimethylammonium chloride) polyelectrolyte film for packaging applications. Int. J. Biol. Macromol. **244**, 125388 (2023). https://doi.org/10.1016/j.ijbiomac.2023.125388
44. J. Yuan, C. Yi, H. Jiang, F. Liu, G.J. Cheng, Direct ink writing of hierarchically porous cellulose/alginate monolithic hydrogel as a highly effective adsorbent for environmental applications. ACS Appl. Polym. Mater. **3**, 699–709 (2021). https://doi.org/10.1021/acsapm.0c01002
45. D. Akın Şahbaz, Gelatin-based hydrogels and ferrogels as smart drug delivery systems: synthesis, characterization and drug release kinetics. Polym. Bull. **81**, 5215–5235 (2024). https://doi.org/10.1007/s00289-023-04963-7
46. A.A. Mir, A.A. Amooey, S. Ghasemi, Adsorption of direct yellow 12 from aqueous solutions by an iron oxide-gelatin nanoadsorbent; kinetic, isotherm and mechanism analysis. J. Clean Prod. **170**, 570–580 (2018). https://doi.org/10.1016/j.jclepro.2017.09.101
47. Y. Ji, X. Yang, Z. Ji, L. Zhu, N. Ma, D. Chen, X. Jia, J. Tang, Y. Cao, DFT-calculated IR spectrum amide I, II, and III band contributions of *N*-methylacetamide fine components. ACS Omega **5**, 8572–8578 (2020). https://doi.org/10.1021/acsomega.9b04421
48. R. Jahanban-Esfahlan, H. Derakhshankhah, B. Haghshenas, B. Massoumi, M. Abbasian, M. Jaymand, A bio-inspired magnetic natural hydrogel containing gelatin and alginate as a drug delivery system for cancer chemotherapy. Int. J. Biol. Macromol. **156**, 438–445 (2020). https://doi.org/10.1016/j.ijbiomac.2020.04.074
49. V.D.A. Pereira, I.S. Ribeiro, H.C.B. Paula, R.C.M. de Paula, R.L. Sommer, R.J.S. Rodriguez, F.O.M.S. Abreu, Chitosan-based hydrogel for magnetic particle coating. React. Funct. Polym. **146**, 104431 (2020). https://doi.org/10.1016/j.reactfunctpolym.2019.104431

50. Y.P. Barragan Medina, V.A. Alvarez, P. Mendoza Zélis, J.S. Gonzalez, Ecofriendly magnetic gels beads based on carboxymethylcellulose and iron oxides for diclofenac adsorption. Discov. Chem. Eng. **4**, 22 (2024). https://doi.org/10.1007/s43938-024-00060-x
51. M. Rasoulzadeh, H. Namazi, Carboxymethyl cellulose/graphene oxide bio-nanocomposite hydrogel beads as anticancer drug carrier agent. Carbohydr. Polym. **168**, 320–326 (2017). https://doi.org/10.1016/j.carbpol.2017.03.014
52. X. Yang, X. Zhang, Y. Ma, Y. Huang, Y. Wang, Y. Chen, Superparamagnetic graphene oxide–Fe_3O_4 nanoparticles hybrid for controlled targeted drug carriers. J. Mater. Chem. **19**, 2710 (2009). https://doi.org/10.1039/b821416f
53. A.C. Ferrari, Raman spectroscopy of graphene and graphite: disorder, electron–phonon coupling, doping and nonadiabatic effects. Solid State Commun. **143**, 47–57 (2007). https://doi.org/10.1016/j.ssc.2007.03.052
54. R. Saito, T. Takeya, T. Kimura, G. Dresselhaus, M.S. Dresselhaus, Raman intensity of single-wall carbon nanotubes. Phys. Rev. B **57**, 4145–4153 (1998). https://doi.org/10.1103/PhysRevB.57.4145
55. A.C. Ferrari, J. Robertson, Resonant Raman spectroscopy of disordered, amorphous, and diamond-like carbon. Phys. Rev. B **64**, 075414 (2001). https://doi.org/10.1103/PhysRevB.64.075414
56. G. Agrawal, S.K. Samal, Raman spectroscopy for advanced polymeric biomaterials. ACS Biomater. Sci. Eng. **4**, 1285–1299 (2018). https://doi.org/10.1021/acsbiomaterials.8b00258
57. P. Tan, Y. Deng, Q. Zhao, Temperature-dependent Raman spectra and anomalous Raman phenomenon of highly oriented pyrolytic graphite. Phys. Rev. B **58**, 5435–5439 (1998). https://doi.org/10.1103/PhysRevB.58.5435
58. P. Tan, C. Hu, J. Dong, W. Shen, B. Zhang, Polarization properties, high-order Raman spectra, and frequency asymmetry between Stokes and anti-Stokes scattering of Raman modes in a graphite whisker. Phys. Rev. B **64**, 214301 (2001). https://doi.org/10.1103/PhysRevB.64.214301
59. M.D. Navarro-Martinez, Tea polyphenol epigallocatechin-3-gallate inhibits ergosterol synthesis by disturbing folic acid metabolism in *Candida albicans*. J. Antimicrob. Chemother. **57**, 1083–1092 (2006). https://doi.org/10.1093/jac/dkl124
60. Y. Ning, J. Ling, C.D. Wu, Synergistic effects of tea catechin epigallocatechin gallate and antimycotics against oral *Candida* species. Arch. Oral Biol. **60**, 1565–1570 (2015). https://doi.org/10.1016/j.archoralbio.2015.07.001
61. J. Guo, X. Song, R. Li, Q. Zhang, S. Zheng, Q. Li, B. Tao, Isolation of a degrading strain of *Fusarium verticillioides* and bioremediation of glyphosate residue. Pestic. Biochem. Physiol. **182**, 105031 (2022). https://doi.org/10.1016/j.pestbp.2021.105031
62. T. Liu, J. Wang, M. Du, Y. Wang, X. Fang, H. Peng, Q. Shi, X. Xie, G. Zhou, The interplays between epigallocatechin-3-gallate (EGCG) and *Aspergillus niger* RAF106 based on metabolism. Fungal Biol. **126**, 727–737 (2022). https://doi.org/10.1016/j.funbio.2022.09.001
63. Z. Zhang, Y. Zhao, J. Yang, J. Guo, J. Li, *Talaromyces cellulolyticus* as a promising candidate for biofilm construction and treatment of textile wastewater. Bioresour. Technol. **340**, 125718 (2021). https://doi.org/10.1016/j.biortech.2021.125718
64. A. Virya, K. Lian, Lithium polyacrylate-polyacrylamide blend as polymer electrolytes for solid-state electrochemical capacitors. Electrochem. Commun. **97**, 77–81 (2018). https://doi.org/10.1016/j.elecom.2018.10.026
65. S. Awasthi, C.P. Pandey, K. Balani, Synergistic role of carbonaceous reinforcements on multi length scale tribology of electrophoretically deposited nickel-boron nitride coatings. Mater. Res. Bull. **99**, 61–72 (2018). https://doi.org/10.1016/j.materresbull.2017.10.030
66. D. Zhang, C. Zhang, J. Liu, Q. Chen, X. Zhu, C. Liang, Carbon-encapsulated metal/metal carbide/metal oxide core-shell nanostructures generated by laser ablation of metals in organic solvents. ACS Appl. Nano Mater. **2**, 28–39 (2019). https://doi.org/10.1021/acsanm.8b01541
67. M. Jakubowski, A. Domke, M. Ratajczak, J. Szczuka, T. Buchwald, Ł Ławniczak, J. Homa, A. Voelkel, M. Sandomierski, Chitosan hydrogel modified with lanthanum as a drug delivery system for epigallocatechin gallate: investigation of hydrogel–drug interaction by FT-IR and

Raman spectroscopy. Spectrochim. Acta A Mol. Biomol. Spectrosc. **297**, 122748 (2023). https://doi.org/10.1016/j.saa.2023.122748
68. S. Awasthi, S. De, S.K. Pandey, Advancement in fabrication and characterization techniques of nanocomposites. ACS Omega **9**, 19756–19769 (2024). https://doi.org/10.1021/acsomega.3c09981
69. W.C. Oliver, G.M. Pharr, Measurement of hardness and elastic modulus by instrumented indentation: advances in understanding and refinements to methodology. J. Mater. Res. **19**, 3–20 (2004)
70. S.S. Sheiko, A.V. Dobrynin, Architectural code for rubber elasticity: from supersoft to superfirm materials. Macromolecules **52**, 7531–7546 (2019). https://doi.org/10.1021/acs.macromol.9b01127
71. M. Shibayama, Structure-mechanical property relationship of tough hydrogels. Soft Matter **8**, 8030 (2012). https://doi.org/10.1039/c2sm25325a
72. H.M. James, E. Guth, Statistical thermodynamics of rubber elasticity. J. Chem. Phys. **21**, 1039–1049 (1953). https://doi.org/10.1063/1.1699106
73. F.T. Wall, P.J. Flory, Statistical thermodynamics of rubber elasticity. J. Chem. Phys. **19**, 1435–1439 (1951). https://doi.org/10.1063/1.1748098
74. P.J. Flory, Theory of elasticity of polymer networks. The effect of local constraints on junctions. J. Chem. Phys. **66**, 5720–5729 (1977). https://doi.org/10.1063/1.433846
75. C. Zhang, B. Wu, Y. Zhou, F. Zhou, W. Liu, Z. Wang, Mussel-inspired hydrogels: from design principles to promising applications. Chem. Soc. Rev. **49**, 3605–3637 (2020). https://doi.org/10.1039/C9CS00849G
76. A. Vedadghavami, F. Minooei, M.H. Mohammadi, S. Khetani, A. Rezaei Kolahchi, S. Mashayekhan, A. Sanati-Nezhad, Manufacturing of hydrogel biomaterials with controlled mechanical properties for tissue engineering applications. Acta Biomater. **62**, 42–63 (2017). https://doi.org/10.1016/j.actbio.2017.07.028
77. Z. Yang, X. Chen, Z. Xu, N. Ji, L. Xiong, Q. Sun, Anti-freezing starch hydrogels with superior mechanical properties and water retention ability for 3D printing. Int. J. Biol. Macromol. **190**, 382–389 (2021). https://doi.org/10.1016/j.ijbiomac.2021.08.235
78. M. Angolkar, S. Paramshetti, R.M. Gahtani, M. Al Shahrani, U. Hani, S. Talath, R.A.M. Osmani, A. Spandana, H.V. Gangadharappa, R. Gundawar, Pioneering a paradigm shift in tissue engineering and regeneration with polysaccharides and proteins-based scaffolds: a comprehensive review. Int. J. Biol. Macromol. **265**, 130643 (2024). https://doi.org/10.1016/j.ijbiomac.2024.130643
79. M.M.H. Rumon, M.S. Rahman, A.A. Akib, M.S. Sohag, M.R.A. Rakib, M.A.R. Khan, F. Yesmin, M.S. Shakil, M.M. Rahman Khan, Progress in hydrogel toughening: addressing structural and crosslinking challenges for biomedical applications. Discov. Mater. **5**, 5 (2025). https://doi.org/10.1007/s43939-025-00178-x
80. A. Rajeswari, S. Gopi, E. Jackcina Stobel Christy, K. Jayaraj, A. Pius, Current research on the blends of chitosan as new biomaterials. INC (2020)
81. J.Y. Sun, X. Zhao, W.R.K. Illeperuma, O. Chaudhuri, K.H. Oh, D.J. Mooney, J.J. Vlassak, Z. Suo, Highly stretchable and tough hydrogels. Nature **489**, 133–136 (2012)
82. S. Awasthi, J.K. Gaur, S.K. Pandey, M.S. Bobji, C. Srivastava, High-strength, strongly bonded nanocomposite hydrogels for cartilage repair. ACS Appl. Mater. Interfaces (2021). https://doi.org/10.1021/acsami.1c05394
83. D. Zhou, W. Wang, W. Ma, Y. Xian, Z. Zhang, Z. Pan, Y. Li, L. Huang, L. Liu, Z. Zheng, H. Liu, D. Wu, Cartilage-adaptive hydrogels via the synergy strategy of protein templating and mechanical training. Adv. Mater. (2025). https://doi.org/10.1002/adma.202414081
84. C.W.H. Rajawasam, O.J. Dodo, M.A.S.N. Weerasinghe, I.O. Raji, S.V. Wanasinghe, D. Konkolewicz, N. De Alwis Watuthanthrige, Educational series: characterizing crosslinked polymer networks. Polym. Chem. **15**, 219–247 (2024). https://doi.org/10.1039/D3PY00914A
85. S. Ladet, L. David, A. Domard, Multi-membrane hydrogels. Nature **452**, 76–79 (2008). https://doi.org/10.1038/nature06619

86. Q. Zhang, Y. Chen, P. Wei, Y. Zhong, C. Chen, J. Cai, Extremely strong and tough chitosan films mediated by unique hydrated chitosan crystal structures. Mater. Today **51**, 27–38 (2021). https://doi.org/10.1016/j.mattod.2021.10.030
87. C. Zhao, P. Zhang, J. Zhou, S. Qi, Y. Yamauchi, R. Shi, R. Fang, Y. Ishida, S. Wang, A.P. Tomsia, M. Liu, L. Jiang, Layered nanocomposites by shear-flow-induced alignment of nanosheets. Nature **580**, 210–215 (2020). https://doi.org/10.1038/s41586-020-2161-8
88. M.T.I. Mredha, Y.Z. Guo, T. Nonoyama, T. Nakajima, T. Kurokawa, J.P. Gong, A facile method to fabricate anisotropic hydrogels with perfectly aligned hierarchical fibrous structures. Adv. Mater. **30** (2018). https://doi.org/10.1002/adma.201704937
89. Y. Yu, S. Wang, H. Yu, X. Liao, W. Feng, Anti-freezing conductive hydrogels with exceptional mechanical properties and stable sensing performance at −30 °C. Mater. Horiz. (2025). https://doi.org/10.1039/D4MH01115E
90. B. Zhu, J. Wu, D. Liu, Y. Yan, X. Yang, Y. Wang, C. Bai, D. Hu, Z. Zhang, P. Jiang, X. Wang, Sculpting mechanical properties of hydrogels by patterning seamlessly interlocked stiff skeleton. Adv. Funct. Mater. **35** (2025). https://doi.org/10.1002/adfm.202417477
91. N. Li, S. Qi, L. Buccoli, P. del Pozo Lorenzale, R.G. De La Cruz, A. Harris, Z.J. Zhang, Multi-scale mechanical properties and enhancement mechanism of cellulose-composited hydrogels. Carbohydr. Polym. **357**, 123421 (2025). https://doi.org/10.1016/j.carbpol.2025.123421
92. Y. Wang, Y. Chen, W. Wang, X. Zheng, S. Chen, S. Wang, F. Wang, L. Wang, Y. Hou, C. Li, Mechanical bionic compression resistant fiber/hydrogel composite artificial heart valve suitable for transcatheter surgery. Compos. B Eng. **296**, 112234 (2025). https://doi.org/10.1016/j.compositesb.2025.112234
93. F. Guo, C. Liu, R. Han, Q. Lu, Y. Bai, R. Yang, D. Niu, X. Zhang, Bio-inspired anisotropic polymeric heart valves exhibiting valve-like mechanical and hemodynamic behavior. Sci. China Mater. **63**, 629–643 (2020). https://doi.org/10.1007/s40843-019-1217-4
94. Q. Zhang, G. Yang, L. Xue, G. Dong, W. Su, M. Cui, Z. Wang, M. Liu, Z. Zhou, X. Zhang, Ultrasoft and biocompatible magnetic-hydrogel-based strain sensors for wireless passive biomechanical monitoring. ACS Nano **16**, 21555–21564 (2022). https://doi.org/10.1021/acsnano.2c10404
95. P. Swiatlowska, J.L. Sanchez-Alonso, P.T. Wright, P. Novak, J. Gorelik, Microtubules regulate cardiomyocyte transversal Young's modulus. Proc. Natl. Acad. Sci. **117**, 2764–2766 (2020). https://doi.org/10.1073/pnas.1917171117
96. W.-C. Yeh, P.-C. Li, Y.-M. Jeng, H.-C. Hsu, P.-L. Kuo, M.-L. Li, P.-M. Yang, P.H. Lee, Elastic modulus measurements of human liver and correlation with pathology. Ultrasound Med. Biol. **28**, 467–474 (2002). https://doi.org/10.1016/S0301-5629(02)00489-1
97. J. Park, S. Choi, A.H. Janardhan, S.-Y. Lee, S. Raut, J. Soares, K. Shin, S. Yang, C. Lee, K.-W. Kang, H.R. Cho, S.J. Kim, P. Seo, W. Hyun, S. Jung, H.-J. Lee, N. Lee, S.H. Choi, M. Sacks, N. Lu, M.E. Josephson, T. Hyeon, D.-H. Kim, H.J. Hwang, Electromechanical cardioplasty using a wrapped elasto-conductive epicardial mesh. Sci. Transl. Med. **8** (2016). https://doi.org/10.1126/scitranslmed.aad8568
98. T. Matsuda, R. Kawakami, R. Namba, T. Nakajima, J.P. Gong, Mechanoresponsive self-growing hydrogels inspired by muscle training. Science **363**(2019), 504–508 (1979). https://doi.org/10.1126/science.aau9533
99. X. Xuan, Y. Li, X. Xu, Z. Pan, Y. Li, Y. Luo, L. Sun, Three-dimensional printable magnetic hydrogels with adjustable stiffness and adhesion for magnetic actuation and magnetic hyperthermia applications. Gels **11**, 67 (2025). https://doi.org/10.3390/gels11010067
100. Z. Gong, G. Zhang, X. Zeng, J. Li, G. Li, W. Huang, R. Sun, C. Wong, High-strength, tough, fatigue resistant, and self-healing hydrogel based on dual physically cross-linked network. ACS Appl. Mater. Interfaces **8**, 24030–24037 (2016). https://doi.org/10.1021/acsami.6b05627
101. N. Yuan, L. Xu, H. Wang, Y. Fu, Z. Zhang, L. Liu, C. Wang, J. Zhao, J. Rong, Dual physically cross-linked double network hydrogels with high mechanical strength, fatigue resistance, notch-insensitivity, and self-healing properties. ACS Appl. Mater. Interfaces **8**, 34034–34044 (2016). https://doi.org/10.1021/acsami.6b12243

102. Y. Yang, X. Wang, F. Yang, L. Wang, D. Wu, Highly elastic and ultratough hybrid ionic–covalent hydrogels with tunable structures and mechanics. Adv. Mater. **30** (2018). https://doi.org/10.1002/adma.201707071
103. J.-Y. Sun, X. Zhao, W.R.K. Illeperuma, O. Chaudhuri, K.H. Oh, D.J. Mooney, J.J. Vlassak, Z. Suo, Highly stretchable and tough hydrogels. Nature **489**, 133–136 (2012). https://doi.org/10.1038/nature11409
104. Q. Chen, L. Zhu, C. Zhao, Q. Wang, J. Zheng, A. Robust, One-pot synthesis of highly mechanical and recoverable double network hydrogels using thermoreversible sol-gel polysaccharide. Adv. Mater. **25**, 4171–4176 (2013). https://doi.org/10.1002/adma.201300817
105. S.D. Sarkar, M.M. Uddin, C.K. Roy, M.J. Hossen, M.I. Sujan, M.S. Azam, Mechanically tough and highly stretchable poly(acrylic acid) hydrogel cross-linked by 2D graphene oxide. RSC Adv. **10**, 10949–10958 (2020). https://doi.org/10.1039/D0RA00678E
106. M.I. Sujan, S.D. Sarkar, S. Sultana, L. Bushra, R. Tareq, C.K. Roy, M.S. Azam, Bi-functional silica nanoparticles for simultaneous enhancement of mechanical strength and swelling capacity of hydrogels. RSC Adv. **10**, 6213–6222 (2020). https://doi.org/10.1039/C9RA09528D
107. M.M.H. Rumon, S.D. Sarkar, M.M. Uddin, M.M. Alam, S.N. Karobi, A. Ayfar, M.S. Azam, C.K. Roy, Graphene oxide based crosslinker for simultaneous enhancement of mechanical toughness and self-healing capability of conventional hydrogels. RSC Adv. **12**, 7453–7463 (2022). https://doi.org/10.1039/D2RA00122E
108. M.J. Hossen, S.D. Sarkar, M.M. Uddin, C.K. Roy, M.S. Azam, Mussel-inspired adhesive nano-filler for strengthening polyacrylamide hydrogel. ChemistrySelect **5**, 8906–8914 (2020). https://doi.org/10.1002/slct.202001632
109. A.R. Shakil, M.L. Begum, M.A.A. Shaikh, S. Sultana, M.S. Rahman, M.M.H. Rumon, C.K. Roy, M.A. Haque, Jute fiber reinforced hydrogel composite for removal of methylene blue dye from water. Dhaka Univ. J. Sci. **70**, 59–64 (2022). https://doi.org/10.3329/dujs.v70i2.62608
110. J. Uchida, B. Soberats, M. Gupta, T. Kato, Advanced functional liquid crystals. Adv. Mater. **34** (2022). https://doi.org/10.1002/adma.202109063
111. T. Kato, J. Uchida, T. Ichikawa, T. Sakamoto, Functional liquid crystals towards the next generation of materials. Angew. Chem. Int. Ed. **57**, 4355–4371 (2018). https://doi.org/10.1002/anie.201711163
112. P.G. de Gennes, Soft matter. Science **256**(1992), 495–497 (1979). https://doi.org/10.1126/science.256.5056.495
113. Z. Zhang, C. Shen, P. Zhang, S. Xu, L. Kong, X. Liang, C. Li, X. Qiu, J. Huang, X. Cui, Fundamental, mechanism and development of hydration lubrication: from bio-inspiration to artificial manufacturing. Adv. Colloid Interface Sci. **327**, 103145 (2024). https://doi.org/10.1016/j.cis.2024.103145
114. G. Zhang, A. Zhu, S. Wang, Q. Chen, B. Liu, J. Zhou, Z. Wu, Stabilizing liquid crystal droplets with hydrogel films and its application in monitoring adenosine triphosphate. Colloids Surf. A Physicochem. Eng. Asp. **654**, 130122 (2022). https://doi.org/10.1016/j.colsurfa.2022.130122
115. R. Qu, T.F. George, G. Li, Development in liquid crystal microcapsules: fabrication, optimization and applications. J. Mater. Chem. C Mater. **10**, 413–432 (2022). https://doi.org/10.1039/D1TC04395A
116. S. Zhang, Y. Qiao, Y. Liu, L. Ma, J. Luo, Molecular behaviors in thin film lubrication—part one: film formation for different polarities of molecules. Friction **7**, 372–387 (2019). https://doi.org/10.1007/s40544-019-0287-1
117. L. Ma, J. Luo, Thin film lubrication in the past 20 years. Friction **4**, 280–302 (2016). https://doi.org/10.1007/s40544-016-0135-5
118. Y. Gong, P. Jing, F. Jing, S. Li, W. Liang, Y. Leng, Liquid crystal/cyclodextrin inclusion complexes embedded hydrogels with dynamic tribological properties. Colloids Surf. A Physicochem. Eng. Asp. **714**, 136602 (2025). https://doi.org/10.1016/j.colsurfa.2025.136602
119. P. Nageshwar, S.W. Wajge, G.L. Dhakar, A.A. Thakre, S. Tripathi, S. Singh, P.K. Maji, C. Das, Fabrication of zinc(II) mediated poly(acrylamide co acrylic acid) hydrogel with thixotropic

and tribological properties. Macromol. Rapid Commun. **46** (2025). https://doi.org/10.1002/marc.202400670
120. A.A. Thakre, A.K. Singh, Determination of work of adhesion of gelatin hydrogels on a glass substrate. Mater. Res. Express **5**, 045302 (2018). https://doi.org/10.1088/2053-1591/aab6a2
121. K.A. Grosch, The relation between the friction and visco-elastic properties of rubber. Proc. R. Soc. Lond. A Math. Phys. Sci. **274**, 21–39 (1963). https://doi.org/10.1098/rspa.1963.0112
122. C. Beddoes, M. Whitehouse, W. Briscoe, B. Su, Hydrogels as a replacement material for damaged articular hyaline cartilage. Materials **9**, 443 (2016). https://doi.org/10.3390/ma9060443
123. W.-L. Duan, L.-N. Zhang, R. Bohara, S. Martin-Saldaña, F. Yang, Y.-Y. Zhao, Y. Xie, Y.-Z. Bu, A. Pandit, Adhesive hydrogels in osteoarthritis: from design to application. Mil. Med. Res. **10**, 4 (2023). https://doi.org/10.1186/s40779-022-00439-3
124. Y. Zhang, C. Tong, Y. Chen, X. Xia, S. Jiang, C. Qiu, J. Pang, Advances in the construction and application of konjac glucomannan-based delivery systems. Int. J. Biol. Macromol. **262**, 129940 (2024). https://doi.org/10.1016/j.ijbiomac.2024.129940
125. X. Pan, Q. Zong, C. Liu, H. Wu, B. Fu, Y. Wang, W. Sun, Y. Zhai, Konjac glucomannan exerts regulatory effects on macrophages and its applications in biomedical engineering. Carbohydr. Polym. **345**, 122571 (2024). https://doi.org/10.1016/j.carbpol.2024.122571
126. D. Kulkarni, V. Agnihotri, S. Bhinge, M. Ban, D. Bari, C.V. Pardeshi, Konjac glucomannan: a functional biopolymer for multifaceted drug delivery applications. Polym. Adv. Technol. **35** (2024). https://doi.org/10.1002/pat.6512
127. P. Wang, Q. Liao, H. Zhang, Polysaccharide-based double-network hydrogels: polysaccharide effect, strengthening mechanisms, and applications. Biomacromolecules **24**, 5479–5510 (2023). https://doi.org/10.1021/acs.biomac.3c00765
128. M. Usuelli, T. Germerdonk, Y. Cao, M. Peydayesh, M. Bagnani, S. Handschin, G. Nyström, R. Mezzenga, Polysaccharide-reinforced amyloid fibril hydrogels and aerogels. Nanoscale **13**, 12534–12545 (2021). https://doi.org/10.1039/D1NR03133C
129. C. Luo, A. Guo, Y. Zhao, X. Sun, A high strength, low friction, and biocompatible hydrogel from PVA, chitosan and sodium alginate for articular cartilage. Carbohydr. Polym. **286**, 119268 (2022). https://doi.org/10.1016/j.carbpol.2022.119268
130. L. Ji, A. Orthmann, L. Cornacchia, J. Peng, G. Sala, E. Scholten, Effect of different molecular characteristics on the lubrication behavior of polysaccharide solutions. Carbohydr. Polym. **297**, 120000 (2022). https://doi.org/10.1016/j.carbpol.2022.120000
131. J. Shao, J. Pu, F. Chen, Y. Liu, J. Song, Konjac glucomannan-based hydrogels with tunable mechanical strength and frictional resistance for biomedical applications. Int. J. Biol. Macromol. **295**, 139612 (2025). https://doi.org/10.1016/j.ijbiomac.2025.139612
132. L. Vukonic, R.M. Nothnagel, F. Groß, C. Bauer, J. Fritz, P. Thurner, C. Gachot, M.R. Ripoll, Cartilage-inspired biocompatible double network PVA based hydrogels: a critical comparison of the role of agarose, chitosan, and silk fibroin additives. Mater. Today Commun. **45**, 112318 (2025). https://doi.org/10.1016/j.mtcomm.2025.112318
133. J. Chen, C. Li, X. Chen, K. Zhou, H. Li, K. Peng, Y. Yang, Y. Dai, B. Huang, In situ swelling of low-friction, high load-bearing self-bending bilayer hydrogels inspired by articular cartilage. Biomed. Mater. **20**, 025018 (2025). https://doi.org/10.1088/1748-605X/ada7b4
134. Y. Li, P. Tian, H. Cao, Y. Wang, X. Zhao, S. Han, C. Wang, Remarkable enhancement of corrosion resistance and tribological properties of chitosan-MXene based hydrogel coating on the surface of Ti6Al4V alloy. Tribol. Int. **192**, 109229 (2024). https://doi.org/10.1016/j.triboint.2023.109229
135. M. Javan Almasi, D. Xiong, Recent progress in bionic hydrogels for articular cartilage: tribological and mechanical characteristics. J. Bionic Eng. **21**, 653–673 (2024). https://doi.org/10.1007/s42235-024-00480-z
136. Y. Meng, J. Cao, Y. Chen, Y. Yu, L. Ye, 3D printing of a poly(vinyl alcohol)-based nanocomposite hydrogel as an artificial cartilage replacement and the improvement mechanism of printing accuracy. J. Mater. Chem. B **8**, 677–690 (2020). https://doi.org/10.1039/C9TB02278C

137. L. Cui, H. Li, C. Gong, J. Huang, D. Xiong, A biomimetic bilayer coating on laser-textured Ti6Al4V alloy with excellent surface wettability and biotribological properties for artificial joints. Ceram. Int. **48**, 26264–26273 (2022). https://doi.org/10.1016/j.ceramint.2022.05.309
138. L. Huang, Z. Li, S. Ma, D. Ye, J. Yang, G. Qin, H. Yin, Q. Chen, Articular cartilage-inspired hybrid double-network hydrogels with a layered structure and low friction properties. ACS Appl. Polym. Mater. **4**, 7634–7644 (2022). https://doi.org/10.1021/acsapm.2c01274
139. Q. Wei, H. Liu, X. Zhao, W. Zhao, R. Xu, S. Ma, F. Zhou, Bio-inspired hydrogel-polymer brush bi-layered coating dramatically boosting the lubrication and wear-resistance. Tribol. Int. **177**, 108000 (2023). https://doi.org/10.1016/j.triboint.2022.108000
140. Y. Han, J. Yang, W. Zhao, H. Wang, Y. Sun, Y. Chen, J. Luo, L. Deng, X. Xu, W. Cui, H. Zhang, Biomimetic injectable hydrogel microspheres with enhanced lubrication and controllable drug release for the treatment of osteoarthritis. Bioact. Mater. **6**, 3596–3607 (2021). https://doi.org/10.1016/j.bioactmat.2021.03.022
141. Q. Chen, S. Liu, Z. Yuan, H. Yang, R. Xie, L. Ren, Construction and tribological properties of biomimetic cartilage-lubricating hydrogels. Gels **8**, 415 (2022). https://doi.org/10.3390/gels8070415
142. K. Zhang, J. Yang, Y. Sun, Y. Wang, J. Liang, J. Luo, W. Cui, L. Deng, X. Xu, B. Wang, H. Zhang, Gelatin-based composite hydrogels with biomimetic lubrication and sustained drug release. Friction **10**, 232–246 (2022). https://doi.org/10.1007/s40544-020-0437-5
143. A.O. Osaheni, P.T. Mather, M.M. Blum, Mechanics and tribology of a zwitterionic polymer blend: Impact of molecular weight. Mater. Sci. Eng. C **111**, 110736 (2020). https://doi.org/10.1016/j.msec.2020.110736
144. Y. Liu, D. Xiong, X. Zhao, A bionic PEEK composite structure with negatively charged surface adsorbing molecular brushes possessing improved biotribological properties for artificial joints. Tribol. Int. **155**, 106808 (2021). https://doi.org/10.1016/j.triboint.2020.106808
145. F. Xiao, J. Tang, X. Huang, W. Kang, G. Zhou, A robust, low swelling, and lipid-lubricated hydrogel for bionic articular cartilage substitute. J. Colloid Interface Sci. **629**, 467–477 (2023). https://doi.org/10.1016/j.jcis.2022.08.146
146. Y. Liu, D. Xiong, A tannic acid-reinforced PEEK-hydrogel composite material with good biotribological and self-healing properties for artificial joints. J. Mater. Chem. B **9**, 8021–8030 (2021). https://doi.org/10.1039/D1TB01357B
147. R. Zhang, W. Zhao, F. Ning, J. Zhen, H. Qiang, Y. Zhang, F. Liu, Z. Jia, Alginate fiber-enhanced poly(vinyl alcohol) hydrogels with superior lubricating property and biocompatibility. Polymers (Basel) **14**, 4063 (2022). https://doi.org/10.3390/polym14194063
148. Q. Chen, X. Zhang, K. Chen, C. Feng, D. Wang, J. Qi, X. Li, X. Zhao, Z. Chai, D. Zhang, Bilayer hydrogels with low friction and high load-bearing capacity by mimicking the oriented hierarchical structure of cartilage. ACS Appl. Mater. Interfaces **14**, 52347–52358 (2022). https://doi.org/10.1021/acsami.2c13641
149. X. Zhao, N. Karthik, D. Xiong, Y. Liu, Bio-inspired surface modification of PEEK through the dual cross-linked hydrogel layers. J. Mech. Behav. Biomed. Mater. **112**, 104032 (2020). https://doi.org/10.1016/j.jmbbm.2020.104032
150. H. Liu, X. Zhao, Y. Zhang, S. Ma, Z. Ma, X. Pei, M. Cai, F. Zhou, Cartilage mimics adaptive lubrication. ACS Appl. Mater. Interfaces **12**, 51114–51121 (2020). https://doi.org/10.1021/acsami.0c15693
151. Q. Chen, X. Zhang, S. Liu, K. Chen, C. Feng, X. Li, J. Qi, Y. Luo, H. Liu, D. Zhang, Cartilage-bone inspired the construction of soft-hard composite material with excellent interfacial binding performance and low friction for artificial joints. Friction **11**, 1177–1193 (2023). https://doi.org/10.1007/s40544-022-0645-2
152. X. Zhang, Z. Lou, X. Yang, Q. Chen, K. Chen, C. Feng, J. Qi, Y. Luo, D. Zhang, Fabrication and characterization of a multilayer hydrogel as a candidate for artificial cartilage. ACS Appl. Polym. Mater. **3**, 5039–5050 (2021). https://doi.org/10.1021/acsapm.1c00807
153. J. Li, L. Gao, R. Xu, S. Ma, Z. Ma, Y. Liu, Y. Wu, L. Feng, M. Cai, F. Zhou, Fibers reinforced composite hydrogels with improved lubrication and load-bearing capacity. Friction **10**, 54–67 (2022). https://doi.org/10.1007/s40544-020-0389-9

154. Z. Ye, H. Lu, G. Chai, C. Wu, J. Chen, L. Lv, Glycerol-modified poly(vinyl alcohol)/ poly(ethylene glycol) self-healing hydrogel for artificial cartilage. Polym. Int. **72**, 27–38 (2023). https://doi.org/10.1002/pi.6444
155. G. Morgese, S.N. Ramakrishna, R. Simic, M. Zenobi-Wong, E.M. Benetti, Hairy and slippery polyoxazoline-based copolymers on model and cartilage surfaces. Biomacromolecules **19**, 680–690 (2018). https://doi.org/10.1021/acs.biomac.7b01828
156. R. Xu, X. Zhao, S. Ma, Z. Ma, R. Wang, M. Cai, F. Zhou, Hydrogen bonding induced enhancement for constructing anisotropic sugarcane composite hydrogels. J. Appl. Polym. Sci. **138** (2021). https://doi.org/10.1002/app.51374
157. L. Cui, H. Li, J. Huang, D. Xiong, Improved biotribological performance of Ti6Al4V alloy through the synergetic effects of porous TiO_2 layer and zwitterionic hydrogel coating. Ceram. Int. **47**, 34970–34978 (2021). https://doi.org/10.1016/j.ceramint.2021.09.038
158. Y. Shi, J. Liu, J. Li, D. Xiong, D. Dini, Improved mechanical and tribological properties of PAAm/PVA hydrogel-Ti6Al4V alloy configuration for cartilage repair. J. Polym. Res. **29**, 515 (2022). https://doi.org/10.1007/s10965-022-03355-1
159. S. Deng, L. Wang, C. Zhao, D. Xiang, H. Li, B. Wang, Z. Li, H. Zhou, Y. Wu, Low coefficient of friction hydrogels with fast self-healing properties inspired by articular cartilage. Colloids Surf. A Physicochem. Eng. Asp. **656**, 130380 (2023). https://doi.org/10.1016/j.colsurfa.2022.130380
160. L. Cui, J. Chen, C. Yan, D. Xiong, Mechanical and biotribological properties of PVA/SB triple-network hydrogel for biomimetic artificial cartilage. J. Bionic Eng. **20**, 1072–1082 (2023). https://doi.org/10.1007/s42235-022-00305-x
161. J. Chen, L. Cui, C. Yan, D. Xiong, Mechanical and tribological study of PVA–pMPDSAH double-network hydrogel prepared by ultraviolet irradiation and freeze-thaw methods for bionic articular cartilage. J. Bionic Eng. **18**, 1192–1201 (2021). https://doi.org/10.1007/s42235-021-00079-8
162. Z. Ye, H. Lu, E. Jia, J. Chen, L. Fu, Organic solvents enhance polyvinyl alcohol/polyethylene glycol self-healing hydrogels for artificial cartilage. Polym. Adv. Technol. **33**, 3455–3469 (2022). https://doi.org/10.1002/pat.5799
163. L. Cui, W. Tong, H. Zhou, C. Yan, J. Chen, D. Xiong, PVA-BA/PEG hydrogel with bilayer structure for biomimetic articular cartilage and investigation of its biotribological and mechanical properties. J. Mater. Sci. **56**, 3935–3946 (2021). https://doi.org/10.1007/s10853-020-05467-9
164. Y. Xi, P.K. Sharma, H.J. Kaper, C.-H. Choi, Tribological properties of micropored poly(2-hydroxyethyl methacrylate) hydrogels in a biomimetic aqueous environment. ACS Appl. Mater. Interfaces **13**, 41473–41484 (2021). https://doi.org/10.1021/acsami.1c13718
165. M. Franco, T. Hari Krishna, A.M. Pillai, A. Rajendra, A.K. Sharma, A comparative study on the corrosion behaviour of hard anodic coatings on AA 6061 obtained using DC and pulsed DC power sources. Acta Metall. Sin. **26**, 647–656 (2013)
166. M. Ebadi, W.J. Basirun, S.Y. Leng, M.R. Mahmoudian, Investigation of corrosion inhibition properties of caffeine on nickel by electrochemical techniques. Int. J. Electrochem. Sci. **7**, 8052–8063 (2012)
167. S. Awasthi, S.K. Pandey, H.J. Shwetha, Nehal, S. Selvaraj, Tough, durable and strongly bonded self-healing cartilage-mimicking noncovalently assembled hydrogel nanostructures: the interplay between experiment and theory. Nanoscale **17**, 1616–1643 (2025). https://doi.org/10.1039/D4NR03322A
168. R. Yan, S. Chen, B. Wang, C. Liu, X. Chen, Magnetic field-oriented conductive decellularized extracellular matrix hydrogel synergizes with electrical stimulation to promote spinal cord injury repair and electrophysiological function restoration. Biomater. Adv. **169**, 214169 (2025). https://doi.org/10.1016/j.bioadv.2024.214169
169. Y. Chen, M. Abe, I. Tomita, Y. Kurashina, Y. Kitamoto, Synthesis and characterization of pH-responsive ferrogels comprising sulfamethazine-based polymer and magnetic nanoparticles for sensing ammonia gas. J. Magn. Magn. Mater. **565**, 170201 (2023). https://doi.org/10.1016/j.jmmm.2022.170201

170. D. Markovic, B. Petrovic, V. Jokanovic, T. Peric, B. Colovic, I. Karadzic, Nanomaterials as scaffolds in bone tissue engineering in dental medicine, in *Nanobiomaterials in Hard Tissue Engineering* (Elsevier, 2016), pp. 413–442. https://doi.org/10.1016/B978-0-323-42862-0.00014-6
171. C. Xu, Y. Zhang, H. Li, Z. Zhang, W. Sheng, S. Zhang, P. Li, X. Zhang, X. Li, H. Lin, H. Duan, S. Guan, L. Wang, Carboxymethylated yeast β-glucan: biological activity screening in zebrafish, sprayable hydrogel preparation, and wound healing study in diabetic mice. Int. J. Biol. Macromol. **285**, 138178 (2025). https://doi.org/10.1016/j.ijbiomac.2024.138178
172. C. Ren, S. Zhang, Z. Zhang, H. Li, W. Sheng, X. Wang, P. Li, X. Zhang, X. Li, H. Lin, H. Duan, S. Guan, L. Wang, Injectable and self-healing carboxymethyl chitosan/carboxymethyl cellulose/marine snail peptide hydrogel for infected wound healing. Int. J. Biol. Macromol. **288**, 138784 (2025). https://doi.org/10.1016/j.ijbiomac.2024.138784
173. E.S. Ko, C. Kim, Y. Choi, K.Y. Lee, 3D printing of self-healing ferrogel prepared from glycol chitosan, oxidized hyaluronate, and iron oxide nanoparticles. Carbohydr. Polym. **245**, 116496 (2020). https://doi.org/10.1016/j.carbpol.2020.116496
174. Z. Liu, J. Liu, X. Cui, X. Wang, L. Zhang, P. Tang, Recent advances on magnetic sensitive hydrogels in tissue engineering. Front. Chem. **8** (2020). https://doi.org/10.3389/fchem.2020.00124
175. T. Carvalho, R. Bártolo, S.N. Pedro, B.F.A. Valente, R.J.B. Pinto, C. Vilela, M.-A. Shahbazi, H.A. Santos, C.S.R. Freire, Injectable nanocomposite hydrogels of gelatin-hyaluronic acid reinforced with hybrid lysozyme nanofibrils-gold nanoparticles for the regeneration of damaged myocardium. ACS Appl. Mater. Interfaces **15**, 25860–25872 (2023). https://doi.org/10.1021/acsami.3c03874
176. R. Li, X. Gong, C. Hong, H. Wang, Y. Chen, K. Tan, X. Liu, F. Wang, An efficient photochemotherapy nanoplatform based on the endogenous biosynthesis of photosensitizer in macrophage-derived extracellular vesicles. Biomaterials **279**, 121234 (2021). https://doi.org/10.1016/j.biomaterials.2021.121234
177. Z.-C. Hu, J.-Q. Lu, T.-W. Zhang, H.-F. Liang, H. Yuan, D.-H. Su, W. Ding, R.-X. Lian, Y.-X. Ge, B. Liang, J. Dong, X.-G. Zhou, L.-B. Jiang, Piezoresistive MXene/silk fibroin nanocomposite hydrogel for accelerating bone regeneration by re-establishing electrical microenvironment. Bioact. Mater. **22**, 1–17 (2023). https://doi.org/10.1016/j.bioactmat.2022.08.025
178. B.Y. Shin, B.G. Cha, J.H. Jeong, J. Kim, Injectable macroporous ferrogel microbeads with a high structural stability for magnetically actuated drug delivery. ACS Appl. Mater. Interfaces **9**, 31372–31380 (2017). https://doi.org/10.1021/acsami.7b06444
179. Y. Lv, F. Cai, Y. He, L. Li, Y. Huang, J. Yang, Y. Zheng, X. Shi, Multi-crosslinked hydrogels with strong wet adhesion, self-healing, antibacterial property, reactive oxygen species scavenging activity, and on-demand removability for seawater-immersed wound healing. Acta Biomater. **159**, 95–110 (2023). https://doi.org/10.1016/j.actbio.2023.01.045
180. M.J. Lee, S.E. Kim, J. Park, G.Y. Ahn, T.H. Yun, I. Choi, H. Kim, S. Choi, Curcumin-loaded biodegradable polyurethane scaffolds modified with gelatin using 3D printing technology for cartilage tissue engineering. Polym. Adv. Technol. **30**, 3083–3090 (2019). https://doi.org/10.1002/pat.4740
181. S. Buvana, J. Charles, V. Kumaravel, S.K. Sadasivam, Biogenic synthesis and characterization of pH-sensitive magnetic PVA/starch/$MnFe_2O_4$ hydrogel for prolonged release of 5-fluorouracil. J. Inorg. Organomet. Polym. Mater. **34**, 5408–5424 (2024). https://doi.org/10.1007/s10904-024-03175-4
182. J. Liu, C. Du, W. Huang, Y. Lei, Injectable smart stimuli-responsive hydrogels: pioneering advancements in biomedical applications. Biomater. Sci. **12**, 8–56 (2024). https://doi.org/10.1039/D3BM01352A
183. J. Hurtaud, C. Delacour, C. Mathevon, V. Forge, Biomimetic hydrogel based on HET-s amyloid fibers for long-term culture of primary hippocampal neurons. Biol. Phys. (2024)
184. G. Mogli, I. Roppolo, A. Chiappone, S. Stassi, Multi-responsive 3D printable organohydrogel for the fabrication of durable and low-hysteresis flexible sensors. Appl. Mater. Today **44**, 102675 (2025). https://doi.org/10.1016/j.apmt.2025.102675

185. F. Ding, L. Zhang, X. Chen, W. Yin, L. Ni, M. Wang, Photothermal nanohybrid hydrogels for biomedical applications. Front. Bioeng. Biotechnol. **10** (2022). https://doi.org/10.3389/fbioe.2022.1066617
186. A. Mohanty, A. Mohapatra, A. Sundaram, P. Sathiyamoorthy, W. Park, S.K. Rajendrakumar, I.-K. Park, Triple-action cancer therapy using laser-activated NO-releasing metallomicellar nanophotosensitizer for pyroptosis-driven immune reprogramming. J. Control. Release **379**, 147–163 (2025). https://doi.org/10.1016/j.jconrel.2025.01.012

Chapter 5
3D, 4D, and Bioprinting of Gels

5.1 3D Printing of Hydrogels and Ferrogels

The quick development of 3D printing technology has completely changed design, development and manufacturing process among various sectors. An age of unmatched customization, shorter lead times, and complex geometries that were previously impossible with conventional production techniques has been brought about by 3D printing in a variety of industries, including aerospace and healthcare [1]. 3D printing has overcome traditional constraints and enabled previously unheard-of levels of creativity by using layers of materials to create 3D items straight from computer blueprints. 3D printing has numerous applications in a variety of industries and materials, including composites. The technique has made it possible to produce intricate, lightweight parts with improved performance attributes for the aerospace industry. It has made it easier for the automobile sector to quickly prototype components and create customized tools as needed. Similar to this, 3D printing has advanced the healthcare industry by creating bioengineered tissues, anatomical models for surgery planning, and implants tailored to each patient [2, 3].

One class of materials has been especially interesting for 3D printing throughout this technological evolution: hydrogels and ferrogels (HG-FGs) [4]. Large volumes of water or biological fluids may be absorbed and retained by the three dimensional network of HG-FGs [5]. Because of this special characteristic, HG-FGs are naturally compatible with biological systems, which makes them a great choice for a variety of applications. Natural, synthetic, hybrid, and chemically and physically crosslinked HG-FGs are among the several varieties of HG-FGs. HG-FGs are a remarkable material for 3D printing because of their high-water content and adjustable mechanical characteristics. HG-FGs may be effectively printed using a variety of 3D printing methods, such as digital light processing, inkjet-based printing [6], 3D printing based on extrusion [7], stereolithography [8], laser-assisted bioprinting [9], etc. Naghieh et al. [10] examined the printability of 3D hydrogel scaffolds, focusing on the influence of hydrogel composition and printing parameters, in order to improve 3D

S. Awasthi, *Smart Hydrogels and Ferrogels*, Materials Horizons: From Nature to Nanomaterials, https://doi.org/10.1007/978-981-95-9077-3_5

printing processes for creating intricate structures used in tissue engineering and other applications. Kalyan and Kumar [11] provided a comprehensive overview of the current status of 3D printing technology in biology and healthcare, as well as insights into the many different uses of 3D printing, including medication delivery, medical devices, and tissue engineering. This adaptability enables advancements in wound healing, drug delivery, tissue engineering, and other areas. The potential for creating biomimetic structures that closely mimic physiological environments is another benefit of hydrogels' capacity to interact with cells and tissues. This would enhance their use in regenerative medicine and disease modeling. The various types of 3D printing methods are schematically depicted in Fig. 5.1. Kapusta and co-workers have examined the use of natural hydrogels with antibacterial properties in biomedicine [12], examining the characteristics, potential applications, and challenges of hydrogels. This work elucidates the application of these hydrogels in various medical fields. The applicability of hydrogels in 3D printing technology was the main emphasis in the study done by Kaliaraj et al. [13]. The authors highlighted the materials' versatility as useful 3D printing materials and their essential function in this innovative manufacturing technique. Zhang et al. [14] published an article that outlines the latest advancements in 3D printing concerning smart hydrogels. Their in-depth research provides valuable knowledge regarding the creation of durable hydrogel materials and their applications, which significantly advances the rapidly evolving field of 3D printing.

HG-FGs-based structures can be designed using a range of 3D printing methods, each with unique benefits and drawbacks. When selecting a 3D printing process for hydrogel-based systems, it is important to carefully consider the trade-offs between resolution, speed, and material compatibility in order to fit the demands of the materials and its intended application [16]. Other 3D printing techniques include stereolithography (SLA), extrusion-based 3D printing, inkjet-based printing, digital light processing (DLP), laser-assisted bioprinting, and others [9, 10]. To create bioinks appropriate for a range of 3D printing processes, Bedell and associates [17] studied human gelatin-based composite hydrogels designed for osteochondral tissue creation. Their research advances the development of new materials and printing techniques for cartilage and joint restoration. There have also been reports on the new field of 3D printing for immobilizing biocatalysts [18, 19]. This work highlights

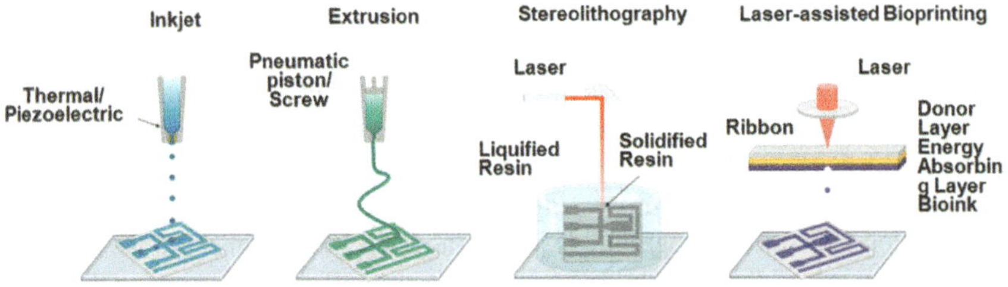

Fig. 5.1 An illustration showing several 3D printing methods. Reproduced with permission from Ref. [15]. Copyright 2024 MDPI. (Open Access)

the significant biotechnology potential of 3D printing, which makes it possible to effectively immobilize enzymes for a variety of uses, including the generation of biofuel and bioprocessing.

5.1.1 Extrusion or Nozzle-Based Printing

The most popular method for creating hydrogel or ferrogel-based scaffolds is extrusion or nozzle-based printing. In this method, melted or viscous gels are squeezed out using a syringe or nozzle, and a three-dimensional object is constructed layer by layer by sequentially extruding materials along a predetermined path that was created by computer modeling [20, 21]. More significantly, before the subsequent layer is deposited, the layers that are printed must be capable of sustaining their own weight and adhering to one another enough. As a result, elements such as the hydrogels' rheological characteristics, solidification temperature, and gel setting mechanism are crucial [22].

Depending on whether a melting process is used, two forms of extrusion or nozzle-based printing may be distinguished. Among the methods based on melting processing are fused deposition modeling, 3D fiber-deposition, multiphase jet solidification technology, and precise extrusion deposition and manufacturing [23, 24]. Due to the utilization of high temperatures, the melt method is often unfavorable from the viewpoint of the printed scaffold's bioactivity. Consequently, efforts were made to substitute dissolution of materials for the melt process; examples of such systems are manufacturing of low-temperature deposition, Mult nozzle deposition manufacturing, pressure-assisted microsyringes, and robocasting. The fused deposition modeling procedure is less flexible than 3D plotting, an extrusion process at low temperatures. One of the most important features of 3D plotting, which was created in 2000 at the Freiburg Materials Research Centre, is its capacity to print soft gel implants with a well-defined internal pore structure. The goal of this method was to satisfy the demands of desktop scaffold creation suitable for tissue engineering applications [25]. This approach compensates for gravitational effects and eliminates the need for support structures during the printing process (Fig. 5.2) [24]. Furthermore, scaffolds may be made from a wider range of natural and synthetic materials, such as pastes and aqueous solutions, than with traditional 3D printing techniques like melting processing. Above all, this 3D plotting technique was created especially for biofunctional processing, allowing for the first time the incorporation of aquatic biosystems like live cells into the scaffold-development process [26].

Viscose hydrogels or ferrogels are first placed into a syringe and then injected via a microneedle that serves as the nozzle for extrusion into a liquid that has the same density as the hydrogel to create scaffolds using this method. Liquid flow can be produced by a volume-driven injection nozzle using a stepper motor or a pneumatic nozzle using filtered air pressure (Fig. 5.2). It is possible to dispense microstrands continuously or microdots intermittently. Through computer-controlled material deposition on a surface, laminar construction is used to create 3D objects.

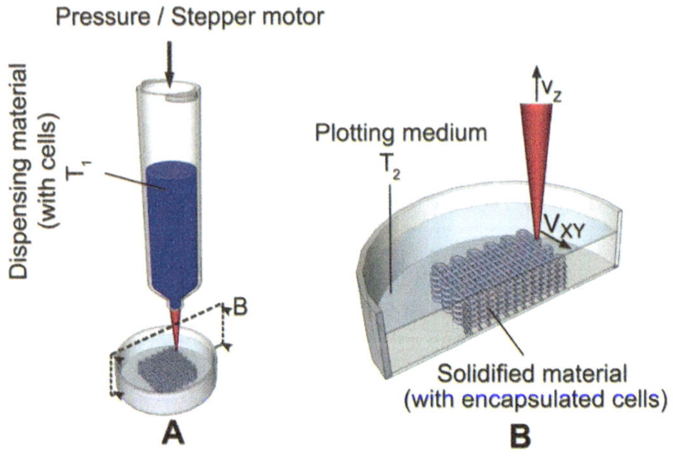

Fig. 5.2 Diagram of the 3D-Bioplotter™ dispensing mechanism. The nozzle of the 3D-Bioplotter™ device operates either by volume-driven injection or pneumatic means. This also exemplifies the broader idea of nozzle-based systems, where material is deposited via a nozzle. The ability to plot into a liquid medium with a corresponding density, so providing buoyancy correction, is a key distinction from previous nozzle-based systems. Reproduced with permission from Ref. [23]. Copyright 2012 Elsevier

The dispensing heads move in three dimensions while the manufacturing platform remains fixed. This buoyancy compensating method is especially advantageous for low viscosity materials. Since heating is not necessary, 3D plotting may also be utilized to build cells and thermally sensitive biocomponents. By changing variables such as the velocity of deposition in the planar direction, the amount of pressure used, the tip diameter, and the hydrogel's viscosity, the microstrand's thickness may be controlled. Curing reactions can be plotted in a co-reactive media or dispensed in two components using a mixing nozzle [23, 25]. However, surface treatment is necessary for scaffolds printed using this technique to make the surface more ideal for cell attachment, since the majority of structures built using 3D plotting have smooth surfaces [25]. The in vitro performance of hydrogel tissue scaffolds created by the freeze-drying technique and those created by 3D plotting was investigated. It was shown that the 3D-plotted scaffolds more effectively promoted osteoblast cell proliferation because they were stronger and less likely to degrade [27]. Recent studies suggested a technique that might be utilized to enhance bone regeneration and address the issue of osteoinduction. It is based on a porous gelatin-alginate scaffold made by 3D plotting [28]. Following the loading of microRNA and a nanodelivery vector as releasable medications within the scaffolds, the combined effects of scaffold degradation and microRNA release were examined in an in vivo animal model. The researchers suggested that since microRNAs are present, this kind of scaffold may encourage osteoblastic differentiation of human bone mesenchymal stem cells and encourage osteogenesis in vivo.

The biocompatibility, low cost, and customizable features of polyacrylamide (PAM) hydrogels have made them highly promising for use in tissue engineering, biosensing, and soft robotics. Direct 3D printing of PAM gel with the required mechanical characteristics is still difficult, though, because of the low viscosity and comparatively quick chemical cross-linking time of the PAM precursor. In order to overcome these obstacles, Xuan et al. [29] used N, N′-methylenebisacrylamide (MBA) to speed up curing and Laponite nanoclay as a nanofiller and rheological/gelling regulator. Combining Laponite nanoclay with MBA resulted in a notable shear-thinning impact in addition to changing the precursor's viscoelastic properties and gelling time. Gel structures with high shape fidelity may be created by optimizing the PAM-Laponite combination for use in traditional extrusion-based 3D printing. The work also demonstrated that the mechanical capabilities of the resulting gel samples may be significantly improved and tuned by the combination of PAM chemical cross-linking, nanoclay physical gelling, and competitive molecular interactions among various gel compositions. In order to prototype 3D gel structures appropriate for biomedical applications, the work showed how to effectively and practically adjust the rheological characteristics of the extrusion precursor.

Collagen material may be 3D printed using freeform reversible embedding of suspended hydrogels (FRESH) technology, with the precise tissue guiding requirements and details required for successful tissue restoration. Even though FRESH has consistently proven to be reliable and successful for extrusion printing, post-print crosslinking techniques can further enhance the mechanical qualities of finished collagen prints. Protein-based materials may be strengthened with ease using photoinitiator-based crosslinking techniques. For collagen materials, the ruthenium and sodium persulfate photoinitiator system ($Ru(bpy)_3$/SPS) was proposed as an efficient crosslinking technique. By integrating extrusion-based 3D printing of type-1 collagen using FRESH technology with Ru(bpy)3/SPS photoinitiated crosslinking procedures, Steiner and the research team [30] described how to improve the strength and endurance of 3D printed collagen structures. To examine the effects of $Ru(bpy)_3$/SPS photoinitiated crosslinking and draw attention to its possible drawbacks, mechanical tests and cell biocompatibility evaluations were conducted. According to these findings, type-1 collagen samples' compressive strength significantly improves with increasing $Ru(bpy)_3$/SPS concentration.

Coffee parchment (CP), one of the neglected by-products of coffee processing, is mostly used for low-value uses like fuel or compost. TEMPO-oxidized nanocellulose (TONCs), which are made from coffee parchment, were investigated by Panyamao et al. [31] for their potential use as a strengthening component in hydrogels (sodium alginate) for the 3D printing of biphasic osteochondral scaffolds (extrusion-based). Using sodium hypochlorite (in varying concentrations of 5–20 mmol/g), TONCs were created by TEMPO-mediated oxidation, producing cellulose nanofibers (TOCNFs: TONC-5, TONC-10) and cellulose nanocrystals (TOCNCs: TONC-15, TONC-20). As seen in Fig. 5.3i, using HSANC inks and the exact printing conditions for the bottom phase (bone-like phase; BS scaffolds), a Petri dish was first printed. An upper phase (cartilage-like phase; CS scaffolds) was then extruded using SANC inks at a speed of 10 mm/s and a pressure of 1 bar. It's interesting to note that the two phases of the biphasic scaffolds, known as BiSANC scaffolds, were designed to be printed using inks that contained the identical kind of TONCs. The cartridge and the printing platform were maintained at a steady 25 °C. The printed structures were thoroughly crosslinked with 1% w/v CaCl2 for 30 minutes, then rinsed with deionized water and freeze-dried. According to rheological analysis, TONC-10-reinforced hydrogels exhibited the greatest yield stress (75.2 Pa), consistency coefficient (323.8 Pa s^n), and printability index (0.929) because of nanofiber entanglement, which guarantees smooth extrusion and filament stability. However, the higher carboxylate content and smaller aspect ratio of TOCNCs, which enhanced ionic crosslinking, hydration, and electrostatic repulsion, allowed for faster structural reformation in TONC-20-reinforced hydrogels, which demonstrated greater thixotropic recovery (64.9%). Following manufacturing, 3D-printed scaffolds reinforced with TONC-20 showed remarkable performance, supporting cell proliferation through improved Young's modulus (227.2 kPa), diffusion-controlled drug release, enhanced water uptake (approximately 1200%), and superior biocompatibility. These findings establish coffee parchment-derived TONCs as sustainable reinforcements for 3D-printed scaffolds by showing that TOCNFs improve printability and extrusion performance while TOCNCs enhance scaffold stability, mechanical strength, and controlled drug release. This study advances environmentally friendly scaffold building for regenerative medicine.

5.1.2 Inkjet-Based Strategy for 3D Printing of HG-FGs

Originally developed as a two-dimensional inkjet printer for text and graphics, this technique has been used as a method of free-form manufacturing to create arrays of proteins and nucleic acids, as well as 3D components [32]. Inkjet is non-contact technique that uses a tiny aperture to jet ink droplets straight to a predetermined spot. There are two types of inkjet printing: drop-on-demand printing and continuous inkjet printing. An electric charge is selectively applied to the droplets as they emerge from the continuous ink stream when the drop break-off mechanism is activated during continuous inkjet printing. Uncharged droplets would proceed directly onto the material to create a picture, but charged droplets are guided into a reservoir for recirculation after passing through the electric field. Drop-on-demand inkjet printing is the preferred method for creating biological structures for soft tissue engineering applications because it eliminates the inherent instability of the recirculation system used in continuous inkjet technology as well as concerns about possible contamination [33]. Moreover, drop-on-demand printing eliminates the need for sophisticated drop charging and deflection gear. Either thermal or piezoelectric force is used to expel liquid droplets. Electrical heating produces pressure pulses that evaporate liquid in thermal inkjet printers.

Then, utilizing air pressure pulses, tiny droplets are released from the nozzle. The temperature may rise to 200–300 °C, which might cause hydrogels or their biocomponents to become denaturalized. When an external voltage is provided, a

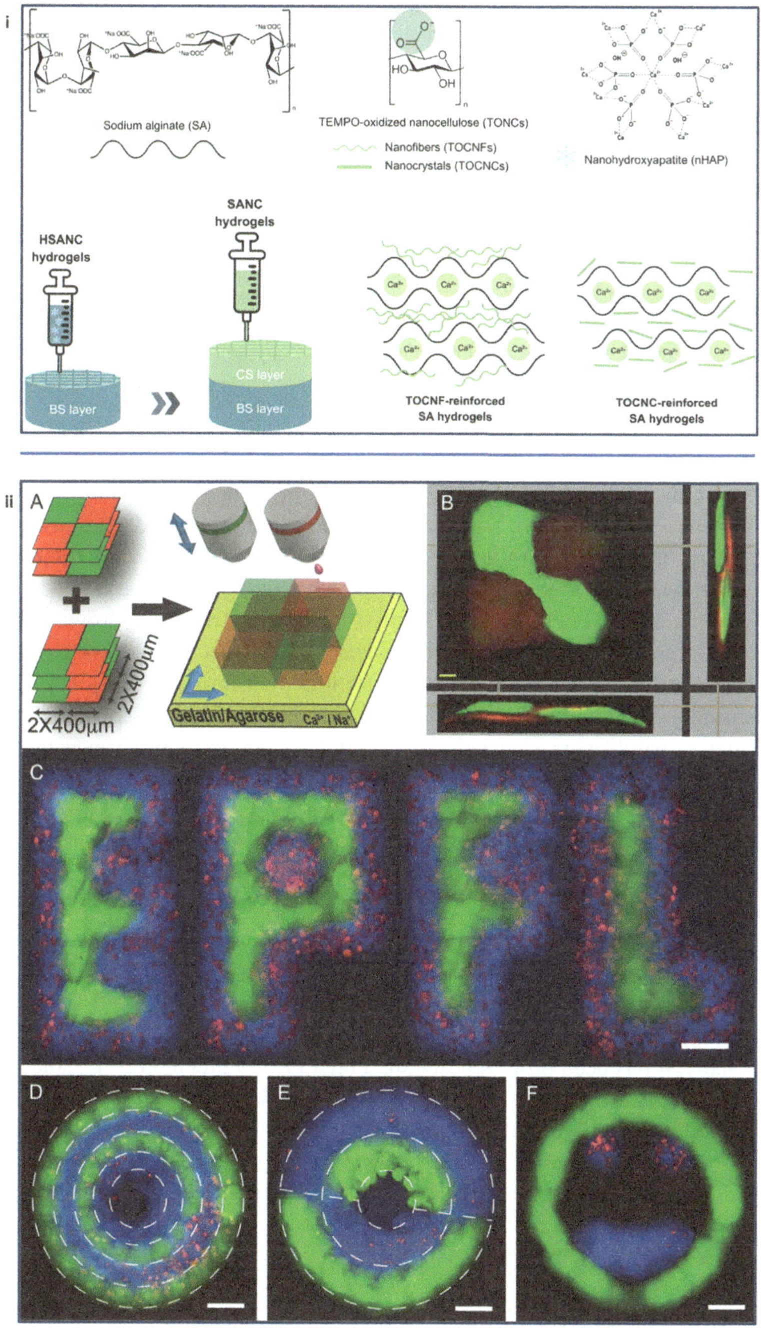
i
Sodium alginate (SA)
TEMPO-oxidized nanocellulose (TONCs)
Nanofibers (TOCNFs)
Nanocrystals (TOCNCs)
Nanohydroxyapatite (nHAP)
HSANC hydrogels
SANC hydrogels
CS layer
BS layer
BS layer
TOCNF-reinforced SA hydrogels
TOCNC-reinforced SA hydrogels
ii
A
B
2X400μm
2X400μm
Gelatin/Agarose
C
D
E
F

◂**Fig. 5.3** **i** The design of TEMPO-oxidized nanocellulose (TONC)-reinforced sodium alginate (SA) hydrogels as printable inks for the 3D printing of two-phase scaffolds (BS: Bone-like phase, CS: Cartilage-like phase). Reproduced with permission from Ref. [31]. Copyright 2025 Elsevier. **ii** Patterning of cell-containing hydrogels using many components. **A** Schematic illustration of a three-dimensional checkerboard made up of the Fluo/Acri (top) and Acri/fluo (bottom) patterns, each of which is overprinted three times. **B** 3D checkerboard bioprint. While the front and side views display the vertical color alternation (scale bar: 100 μm), the top view depicts the single-layer fashion. To improve contrast, Acri-ink was turned red. (**C**–**F**) Printing patterns using tomato NIH 3T3 fibroblasts (red) using Fluo-ink (green) and Acri-ink (blue). **C** The university logo, with a single line for the outside stroke and double line letters for the interior area. **D** Fluo-ink and modified Acri-ink were used twice in succession to print concentric circles. **E** Two-lined circles were produced by alternating the two inks in the partial circles design. The blueprints (in **D**, **E**) are indicated by white dashed lines. **F** "Smiley face" to illustrate our method's adaptability (**C**–**F**: scale bar 200 μm). Reproduced with permission from Ref. [38]. Copyright 2018 Springer Nature. (Open Access)

piezoelectric actuator generates pressure, which is subsequently employed to expel droplets from the nozzle in piezoelectric inkjet printers [33]. It should be noted that droplets with diameters ranging from 15 to several hundred microns may be printed using drop-on-demand and continuous inkjet technologies [33]. Inkjet-based bioprinting, which is based on the traditional inkjet printing method, can accurately apply computer-controlled picoliter droplets of "bioink" onto a hydrogel substrate [34]. In contrast to thermal inkjet printers, the piezoelectric method for bioprinting allows droplets to stay oriented with consistent and uniform size [34]. Living cells may be safely ejected onto culture disks, according to research by Nakamura et al. [35]. Later, research using a piezoelectric technique was conducted to investigate how cells behaved when subjected to the mechanical stresses related to the usage of inkjet drop-on-demand printing systems [36]. Additionally, cell survival was unaffected by the rising time; however, it was somewhat impacted by the pulse's amplitude (from 98 to 94%). Cell vitality was also examined after printing, and the printing procedure had no effect on the cells that survived; there was no discernible difference in cell survival between pulse rise time and amplitude. However, as shown in the work by Saunders et al., inkjet printing requires that cell suspensions be stable for a period of minutes throughout the printing process. They discovered that the printing performance was impacted by some cell aggregation or sedimentation after about 20 minutes [36]. Additionally, it has been proposed that the inkjet printing method is the sole way to create bioink that has a viscosity of 10 mPa/s [13]. The size of the deposited droplets may be regulated between 1 and 300 picoliters, with deposition speeds varying from 1 to 10,000 droplets per second [37]. However, before the subsequent layer of droplets is placed, the droplets must solidify into their predetermined shape. Therefore, it has been proposed that in order to use inkjet printing to make 3D hydrogel structures with excellent resolution, characteristics like quick gelation and the distributed droplets' capacity to maintain a 3D structure are essential [38].

The foundation of reactive inkjet printing is the discharge of ink droplets onto a substrate by drop-on-demand using piezoelectric actuators. Reactive inkjet printing, alternatively, changes the printing process from merely a deposition technique to one that allows precise small-scale reactions by adding tiny droplets of one reactant to another [39, 40]. The fundamental concept of reactive inkjet printing is the combination of a droplet of one reactant with a droplet of a second reaction using an inkjet printer [39]. Reactive inkjet printing comes in two varieties: single reactive inkjet printing, in which the inkjet printer deposits a reactant on top of another reactant deposited using a different method, and full reactive inkjet printing, in which the printer can dispense multiple types of ink, allowing the deposition of two or more reactants [39]. Reactive inkjet printing has been used in studies to investigate the creation of hydrogels [40, 41]. Duffy et al. employed alternating ink layers to create a unique surface pattern based on the immediate formation of ionic bonds between polymers with opposing charges in their effort to print polymer hydrogels based on natural peptides, gellan gum, and poly-ε-lysine [40]. Teo et al.'s study explored an approach that concentrated on the collision between the crosslinker and gelator prior to their impact on the substrate surface to create a situation where hydrogel synthesis and patterning may occur in a single process [41]. Since the slow diffusion of the crosslinking species through the gel results in very slow vertical printing rates, this might be problematic for hydrogels. Cell death and gel dehydration may result from the simultaneous deposition of both. The introduction of reactive inkjet printing greatly accelerated vertical printing rates. It has been shown that the droplet collision reduces the time required for gelation to begin and improves the definition of printed patterns by increasing advective mixing prior to contact.

In a recent work carried out by Lee et al. [42], it was presented that sonochemical treatment reduces the length of polymer chains without chemically destroying the methacryloyl groups, hence controlling the viscoelasticity of a bioink based on gelatin methacryloyl (GelMA). A piezo-axial vibrator is used to assess the rheological characteristics of treated GelMA inks throughout a broad frequency range of 10–10,000 Hz. The maximum printed polymer concentration may be successfully raised from 3 to 10% using this method. After crosslinking, the microstructure and mechanical characteristics of GelMA hydrogel structures are efficiently controlled by the sonochemical treatment, which also preserves the fluid properties within the printable range. Fibroblasts may proliferate on GelMA hydrogels as a result of the mechanical qualities being controlled. Using high-resolution inkjet printing, a multilevel hydrogel structure with 3D cells and layers with varying physical characteristics is created. By extending the range of suitable bioinks, the sonochemical treatment offered a new avenue for inkjet bioprinting to create microarchitectures with different physical characteristics.

In another work, Negro et al. [38] presented the 3D hydrogel patterning using inkjet technology in accordance with somewhat intricate blueprints, achieved by carefully optimizing the printing parameters for two distinct bioink formulations. High-resolution, multi-component live constructions were produced by 3D printing bioinks with live cells in them. Lastly, we present a method for inkjet printing perfusable channels using sacrificial materials to enhance the long-term cultures of larger samples. The theoretically straightforward pattern of a 3D checkerboard (Fig. 5.3ii A, B) was first selected to evaluate the quality of confocal microscopy-based 3D patterning. Even with cells in Acri-ink (red dots), the printouts and relative blueprints (white dashed lines) in Fig. 5.3ii (C–F) demonstrate excellent multi-component patterning accuracy. Notably, our printing strategy's adaptability enabled us to create and pattern a large range of 3D designs with an accuracy of around 100 μm that were several centimeters long and millimeters high in subsequent rounds (Table 5.1).

5.1.3 Printing via Laser or Stereolithography (SLA)

Stereolithography-based 3D printing is a cutting-edge additive manufacturing technique that creates intricate three-dimensional objects using light and hydrogel materials. This approach extends the basic principles of traditional stereolithography by modifying them to account for the unique properties of hydrogels, which are often networks of water-based polymers. The method begins with a hydrogel resin that may include crosslinkers, photo-initiators, and other additives. This method hardens specific regions of the hydrogel resin using a precisely controlled light source, usually a laser or projector. As each layer hardens, the construction platform progressively descends, allowing the 3D object to be built piece by piece. Because of its exceptional accuracy and intricacy, stereolithography is well-suited for use in tissue engineering, biomedical devices, and microfluidic systems. After printing, further steps like cleaning and curing could be required to increase structural stability. Using stereolithography in 3D printing to create hydrogels with unique properties and uses has been the subject of several investigations. Kalossaka and associates [58] explored the development of 3D hydrogel nanocomposites, showcasing the technique's potential for designing complex structures. Karakurt et al. [59] used SLA 3D printing to examine the controlled release characteristics of hydrogels loaded with ascorbic acid. The paper provides clarification on the use of SLA in drug delivery systems.

Additionally, the effects of SLA 3D printing on PEGDMA hydrogel characteristics have been discussed, providing information on the interplay between printing and hydrogel properties [60]. Magalhães et al. [61] examined the use of low-cost stereolithography technology to create 3D hydrogel constructs, which might have applications in biomaterials and biomedicine. Alketbi et al. [62] have examined how PEGDA photopolymerization in micro-stereolithography affects 3D-printed hydrogel materials, focusing on structural integrity and swelling behavior. With an emphasis on attaining a high mechanical tunability and self-weldability, Sun et al. [63] introduced a new method for 3D printing using stereolithography for more functional and adaptable hydrogels. Table 5.2 lists some 3D-printed hydrogels and ferrogels based on stereolithography for a range of uses. Besides the ability of stereolithography-based 3D printing for HG-FGs to produce intricate, high-resolution structures, there are drawbacks to its use, including a lack of material possibilities and the prerequisite for specialist apparatus. Nonetheless, it remains a powerful tool in the area of hydrogel-based additive manufacturing, enabling the development of complex biomimetic structures for a range of applications.

Risangud et al. [64] used stereolithography to create the self-healing hydrogels (oxidized dextran and carboxymethyl chitosan) for use in biomedical applications. The study introduced a new technique for 3D-printed self-healing hydrogels using

Table 5.1 Types of 3D-printed hydrogels that are based on extrusion and inkjet, as well as the many uses for each

Material composition	3D printing technique	Application or purpose	References
Laponite® with oxidized alginate and gelatin	Extrusion-based bioprinting	Promotes integration of natural hydrogels in advanced fabrication methods	[43]
Cellulose-based derivative	Extrusion printing	Acts as an eco-friendly, degradable structural support	[19]
Gelatin and oxidized nanocellulose	Extrusion-based printing	Well-suited for scaffold fabrication in tissue engineering	[44]
GelMA with Laponite nanoclay	Extrusion bioprinting	Facilitates bone tissue repair using high nanoclay content	[45]
Pre-gelled alginate	Extrusion 3D printing	Engineered for high-performance, printable hydrogel systems	[46]
Light-sensitive polypeptide hydrogels	Extrusion-based printing	Enables construction of stable structures responsive to light	[47]
Cellulose-based hydrogel	Extrusion 3D printing	Supports framework creation through solution-based extrusion	[48]
Composite hydrogel bioinks	Extrusion bioprinting	Offers broad applicability for diverse bioprinting uses	[49]
Chitosan-based hydrogels	Simulation-based extrusion printing	Enables digital process optimization for hydrogel printing	[56]
Itraconazole-loaded hydrogel	Inkjet printing	Targets localized drug release in eye treatments	[50]
Modified hydrogel bioinks	Thermal inkjet bioprinting	Improves printability using heat and chemical treatment	[51]
Hyaluronic acid-based hydrogel	Light-based printing techniques	Allows fine microstructure creation through comparative methods	[52]
Generic hydrogel formulations	Inkjet 3D printing	Builds cell-embedded structures for tissue mimicry	[38]
General-purpose hydrogel	Microreactive inkjet printing	Enables formation of tiny, self-supported 3D features	[53]
Star-shaped copolymer hydrogel	3D inkjet technique	Provides structural stability through metal-ion-induced crosslinking	[54]
Alginate–gelatin blend hydrogel	Inkjet method	Customizes biological and mechanical traits for regenerative purposes	[55]
Layered hydrogel systems	Inkjet-spray hybrid printing	Enables sequential fabrication of multifunctional hydrogel layers	[56]

(continued)

Table 5.1 (continued)

Material composition	3D printing technique	Application or purpose	References
Layered hydrogel constructs	Ionic inkjet method	Achieves shape-changing properties via patterned surface control	[57]
Poly-ε-lysine with gellan gum	Reactive inkjet printing	Produces bioprinted corneal tissues for eye-related applications	[40]

light-assisted 3D printing, which relies on Schiff-base and covalent bonding formations. These bonding resulted from the reaction of amine and aldehyde functional groups and the photopolymerization of methacrylate groups. First, two distinct polymers, carboxymethyl chitosan (CMCs) and dextran, were changed to create methacrylate-modified carboxymethyl chitosan (CMCs-MA) and oxidized dextran (OD). Depending on the polymer amount and level of oxidation, the generated hydrogels' mechanical performance, swelling ratio, microstructure, and biological reactions show substantial variations. The compressive moduli ranged from 14.31 to 26.20 kPa. Chondrocytes cultured with various hydrogel formulations showed distinct cell shapes and adhesion modifications due to the interaction between the mechanical and biochemical properties of the hydrogel. The present study developed a technique for producing 3D-printed self-healing hydrogels with changeable stiffness, which have significant potential for biomedical applications and may be used to modulate chondrocyte form.

The size of nano-Fe_3O_4 has a significant effect on 3D printing properties in the area of stereolithography 3D printing of the hydrogel (photothermal responsive) because of sedimentation; nevertheless, there is not much relevant research. Recent research employed poly(N-isopropylacrylamide-acrylamide)/nano-Fe_3O_4[P(NIPAM-AM)/nano-Fe_3O_4] photoresist to develop a photothermal responsive hybrid hydrogel utilizing digital light processing (DLP), a popular stereolithography 3D printing method [65]. Figure 5.4A shows the P(NIPAM-AM)/nano-Fe_3O_4 gel photoresist fabrication and method of 3D printing. Nano-Fe_3O_4 (0.80 g) was first dissolved using ultrasonic oscillation for 45 minutes in a solution of ethanol (1.10 mL) and DMSO (5.64 mL) as part of a routine preparation procedure followed by using NIPAM and AM as monomers, MBAA as the crosslinker, Omnirad 819 as the photoinitiator, and glycerol as the thickener, a black gel photoresist with 3.11 weight percent nano-Fe_3O_4 was created. To investigate the impact of nano-Fe_3O_4 sizes on photocuring 3D printing, gel photoresist was made using nano-Fe_3O_4(50) and nano-Fe_3O_4(10) with varying concentrations. Furthermore, using gel photoresist with mixed nano-Fe_3O_4, a biomimetic hydrogel octopus was 3D printed. The octopus tentacle successfully waved in reaction to near-infrared light.

Table 5.2 Hydrogels fabricated using stereolithography and digital light processing with their diverse applications

Hydrogel type	Purpose/applications	Technique type	Citations
Nanocellulose/ PEGDA	Engineering 3D cell culture scaffolds for regenerative medicine	Stereolithography-based 3D printing	[83]
Nanocomposite hydrogels	Construction of vascular-mimicking hydrogel structures	Stereolithography-based 3D printing	[58]
Ascorbic acid-enriched hydrogels	Drug delivery through precisely fabricated 3D-printed hydrogel matrices	Stereolithography-based 3D printing	[59]
PEGDMA hydrogels	Studying how 3D printing modifies mechanical and structural hydrogel properties	Stereolithography-based 3D printing	[60]
Hydrogel-based structures	Economical fabrication of hydrogel architectures using additive manufacturing	Stereolithography-based 3D printing	[61]
PEGDA hydrogels	Investigating photopolymerization parameters in micro-stereolithography	Stereolithography-based 3D printing	[62]
Tunable mechanical hydrogels	Designing hydrogels with customizable stiffness and self-repairing characteristics	Stereolithography-based 3D printing	[63]
Hierarchical hydrogels	Creation of hydrogels with multiscale structural complexity	3D printing based on DLP	[84]
Supramolecular hydrogels	Fabricating high-impact absorbing materials with complex internal frameworks	3D printing based on DLP	[85]
Cellulose hydrogel	Application in flexible sensors for mechanical strain detection	3D printing based on DLP	[86]
Cellulose-based hydrogels	Fully cellulose-derived hydrogel printing via digital light projection	3D printing based on DLP	[87]
Antimicrobial hydrogel	Eco-friendly hydrogels with bactericidal properties using nanosphere technology	3D printing based on DLP	[88]
Silk fibroin (A. Pernyi) bioinks	Utilizing silk-derived bioinks in light-based 3D printing	DLP 3D printing	[79]
Double-network hydrogels	Producing durable hydrogel networks with enhanced toughness for 3D printing	3D printing based on DLP	[80]
PEDOT-based photopolymerizable inks	Creating biosensor components via conductive polymer inks and light curing	DLP Photopolymerization	[81]

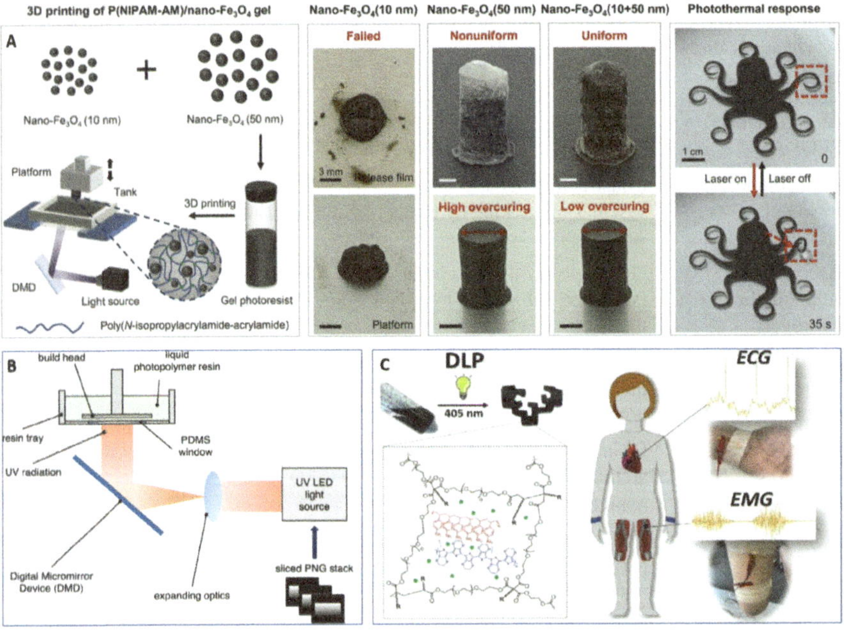

Fig. 5.4 **A** Diagrammatic representation of the DLP 3D-printed P(NIPAM-AM)/nano-Fe_3O_4 gel manufacturing process, showing the use of gel photoresist with single nano-Fe_3O_4(10) and (50) to 3D-printed samples. Pictures of the cylinder sample after freeze-drying, pictures of the cylinder sample after 3D printing, and pictures of the printing residues on the release film and platform using the P(NIPAM-AM)/nano-Fe_3O_4(10) gel photoresist under exposure periods of 15 s. Octopus 3D printing model design diagram. Images showing the biomimetic movement of the octopus tentacle powered by an infrared laser, with the laser switched on and off. Reproduced with permission from Ref. [65]. Copyright 2024 Elsevier. **B** Diagram of a 3D printer using digital light processing (DLP). Reproduced with permission from Ref. [82]. Copyright 2024 Springer Nature. **C** Schematic representation of DLP printed hydrogel for human electrocardiography (ECG) and electromyography (EMG) recordings demonstrate longer-term activity, up to two weeks, and improved detection signals. Reproduced with permission from Ref. [81]. Copyright 2022 American Chemical Society. (Open Access)

5.1.4 Digital Light Processing (DLP) Method

A vat photopolymerization technique called digital light processing may create 3D hydrogels with more geometrical complexity and better precision without significantly affecting printing time due to spatial resolution. Light projection causes localized solidification of a liquid formulation, which allows for the layer-by-layer creation of accurate, self-standing three-dimensional structures with precise spatiotemporal control [66].

First employed by Lu et al., digital light processing is sometimes referred to as digital micromirror device-based stereolithography. It is a simple and rapid layer-by-layer microstereolithography system that consists of a normal computer projector, a digital micromirror masking device, and an ultraviolet light source. A 3D printing system that combines digital light processing (DLP) and photopolymerization for manufacturing is designed in Fig. 5.4B; [67, 68]. Before transferring the specifically created PNG image to the expanding optics to create a layer of the 3D component, the UV LED light source projects the picture onto the 2D plane in accordance with the printing parameters. These expanding optics evenly disperse UV light into the digital micromirror device (DMD) to direct the incoming light distribution toward the liquid photopolymer resin in the resin tray (vat) via the PDMS window. The next layer is printed by the thickness of the previous layer once the exposed liquid polymer dries and sticks to the build head. These processes are carried out repeatedly until the 3D model is completely produced. It makes it possible to create intricate interior structures and precisely distribute biological components in space inside a single scaffold [69]. In contrast to developing the 3D microstructure point-by-point or using a molded structure with thermal lamination, they continued, digital light processing is quite rapid [70, 71]. Hydrogels made from photocrosslinkable PEG diacrylates were used to test the feasibility of using this method to create scaffolds that could promote cell proliferation and differentiation. Each layer or partial layer was created from different controlled-release microparticles by modifying the biofactors or controlled-release particles inside the polymerizable resin. This produced environments that were spatially dispersed and had microsize resolution. Furthermore, intricate and accurate interior structures, including pore size and form, may be produced using a digital mirror device. Cells obtained from mouse bone marrow were effectively planted or encapsulated on fibronectin-functionalized patterned scaffolds.

Subsequent research by Kim et al. also indicated that cells were not harmed during the printing process because of the precise and quick printing speed, and in vitro measurements showed that cells multiplied and dispersed uniformly over the hydrogel [72]. Additionally, the in vivo transplantation of 3D-printed hydrogels employing this technology into a rabbit model with a partially damaged trachea demonstrated the possibility for utilizing digital light processing bioprinting in tissue creation [73]. Notwithstanding the high cost of production, the quick printing speed of 3D printing based on digital light processing has opened up possibilities and allowed for research and development in fields like tissue regeneration [74, 75], drug delivery systems [76], artificial skin models [77], and microfluidics [78]. Koffler et al.'s [74] effort attempted to produce a complicated structure of the central nervous system using a microscale continuous projection printing technology. The researchers claim that this technique can be used to create a 3D biomimetic hydrogel scaffold based on PEG diacrylate and gelatin methacrylate that is as large as the mouse spinal cord in less than two seconds. The scaffold can then be scaled to fit lesion geometries and human spinal cord diameters.

Using easily accessible ingredients and a widely available DLP printer, Caproili and associates [66] introduced a novel hydrogel that can be 3D-printed and has the ability to cure itself. Following polymerization, this linear polymer is effectively encased in the cross-linked matrix. The photocurable ink formulation was made by combining an aqueous solution of unmodified, non-crosslinked poly (vinyl alcohol) (PVA) with acrylic acid (AAc), a water-soluble photoinitiator based on diphenyl

(2,4,6-trimethylbenzoyl)phosphine oxide (TPO), and the cross-linking agent poly (ethylene glycol) diacrylate (PEGDA). Antheraea pernyi silk fibroin bioinks were tested by Zhang et al. [79] for DLP 3D printing, creating new avenues for regenerative medicine and bioprinting. Another important progress was done by Xiang et al. [80], who employed DLP to create 3D-printed hydrogels (double network with enhanced toughness), which helped to create hydrogel materials with high mechanical strength. The use of PEDOT-based photopolymerizable inks for biosensing via DLP 3D printing was investigated by Lopez-Larrea et al. [81], indicating the possibility of improved biosensor development (Fig. 5.4C). Table 5.2 lists many DLP-based 3D-printed hydrogels for a range of uses.

5.2 4D Printing of Hydrogels and Ferrogels

The advent of 4D printing, which outperformed its predecessor, 3D printing, has significantly advanced the area of additive manufacturing. This innovative technique makes use of materials that undergo form and property changes over time, allowing things to react and adjust to certain stimuli [89]. A thorough study is necessary to close this knowledge gap and identify new prospects in this emerging field, since, despite tremendous developments, the incorporation of stimuli-responsive hydrogels into 4D printing for biomedical applications is still little understood. Shape-changing programming is at the core of the 4D printing idea. This entails the careful and exact construction and encoding of structures or materials that are capable of changing their shape in response to certain inputs [90–92]. The functional breadth of the two technologies is one obvious difference between them. While 4D printing uses materials that can change after manufacturing, 3D printing uses a layer-by-layer method to assemble static things carefully [93].

Since its introduction in 2013, the industry of 4D printing has seen substantial advancements and innovations [89]. In order to produce objects that can dynamically change their functionality, texture, and shape over time, researchers are constantly experimenting with different methods and materials. The ability to convert intangible ideas into flexible prototypes has developed innovations in a variety of sectors, from traditional manufacturing to healthcare [91, 92, 94, 95]. The use of biological materials and cell-based treatments to replace or repair damaged tissues and organs is the core of the tissue engineering discipline [96]. One of the main objectives in this discipline is to create synthetic scaffolds that can effectively replace damaged tissues [97]. Effective tissue restoration is made possible by materials with bio-responsive and bio-mimetic qualities [91]. The advent of 4D printing has been a advanced technology in this regard since it can produce complex scaffolds that exhibit controlled changes in characteristics over a predetermined time frame. The ability of 4D printing to create small devices, which lowers the size of surgical incisions, is a key advantage of using technology in medical settings. After being surgically implanted, these devices may change to flawlessly suit the body's architecture and adapt to the complex complexity of the human body [98, 99].

In addition to tissue engineering, 4D printing has become a major study topic, especially in the area of controlled drug delivery systems. This cutting-edge technology opens up new opportunities in the pharmaceutical industry by allowing the sequential delivery of various drugs. The accomplishment is the accurate placement of medicinal substances in specific areas of a delivery system. Furthermore, it is important to note that 4D printed devices may be deliberately designed to allow for the regulated release of drugs in response to particular stimuli [100]. The use of 4D printing in the creation of smart materials offers a significant advantage in the field of medical diagnostics. It is possible to integrate these elements into biological systems, improving their sensing skills [101].

Because of their special qualities, hydrogels and ferrogels have immense importance for 4D bioprinting [102]. Intricate polymeric networks with cross-links connecting them in a three-dimensional structure are known as hydrogels. Their innate hydrophilic and water-insoluble qualities are what distinguish them and help to explain their unique "soft" quality. Because of their natural softness, hydrogels usually exhibit a range of stiffness and pliability, although their mechanical qualities are often poorer [103]. Both natural and manufactured hydrogels fall into one of two primary groups. On the other hand, it is important to remember that natural hydrogels are made of a variety of constituents, like collagen, hyaluronan, chitosan, alginate, fibrin, and agarose [104]. One class of hydrogels, referred to as "smart hydrogels," demonstrates the ability to alter their properties in response to specific stimuli (such as thermo-responsive, pH-responsive, magneto-responsive, and moisture-responsive) [104].

With a special emphasis on hydrogels and their temporal response to biological stimuli, this chapter offers a thorough examination of stimulus-responsive materials. The goal is to give a thorough information of 4D printing's potential in producing advanced and adaptable products, as well as its substantial influence on biomedical advancements, by thoroughly examining the complexities surrounding this technology. The purpose is to investigate the complex interaction between stimuli-responsive hydrogels and 4D printing. It aims to highlight the possibility of creating potential biological therapies that can easily adapt to the shifting needs of the healthcare industry.

5.2.1 Thermo-Responsive

Particularly in the medical field, thermo-responsive or thermosensitive hydrogels have garnered a lot of interest. Several techniques have been developed to regulate the temperature differential from room temperature in vivo, based on the widely recognized average normal body or physiological temperature of 37 °C [104]. A thermo-responsive hydrogel's phase transition characteristics will cause a reaction that may result in a shift in its solvent affinity once it reaches a particular temperature [105]. This is referred to as the critical solution temperature. The chains expand or contract in response to slight temperature variations around a critical solution temperature, which alters the hydrophilic and hydrophobic contacts between the polymer network and the aqueous medium [106]. Additionally, there are two types of thermo-responsive hydrogels: positively and negatively responsive systems. These systems are easily recognized by their respective lower critical solution temperatures (LCST) and upper critical solution temperatures (UCST) [104].

Poly(N-isopropylacrylamide) (P(NIPAAm)) hydrogel is among the most extensively used and researched temperature-responsive hydrogels. Its LCST response at about 32 °C in an aqueous medium is caused by the equilibrium between the hydrophilic and hydrophobic groups as well as the enthalpic and entropic efforts required to solvate these groups [104, 107]. Above the lower critical solution temperature (hydrophobic state), P(NIPAAm) hydrogel displays a collapsed globular form; below the lower critical solution temperature (hydrophilic state), it displays a random coil structure. Furthermore, this hydrogel is inflated below the lower critical solution temperature and absorbs water; yet, it will undergo a sharp and abrupt volume decrease when the medium's temperature climbs above the lower critical solution temperature [107].

P(NIPAAm) hydrogels are mostly made via methods like lithography and molding, which are essentially limited to a 2D area, despite their significant research in tissue engineering and drug administration [108]. In 2018, an effort was made to use high-resolution projection microstereolithography to 3D print P(NIPAAm) hydrogel. The authors claimed that by modifying the manufacturing process parameters and the composition of the polymer resin, the 3D-printed P(NIPAAm) hydrogel's temperature-dependent deformation may be controlled [108].

One of the most widely used 4D printing techniques is micro-extrusion-based 3D printing because of its ease of use and ability to handle a variety of ink ingredients [109]. A hydrogel comprising sodium alginate, methylcellulose, and P(NIPAAm) precursors was created, and its possible use as a material for wound dressings was investigated. Along with acting as P(NIPAAm)'s long-chain crosslinker, PEG diacrylate was introduced as a hydrophilic co-monomer. After that, a combination of 2-phenoxyethanol and octenidine dihydrochloride was added to the micro-extruded hydrogel, making it antibacterial. In addition, the scientists noted that the temperature-stimulated polymer exhibited shape-morphing behavior, controlled swelling ratio, and in vitro drug diffusion, in addition to having local temperature sensing capabilities [110]. Furthermore, it has been suggested that combining microfluidic devices with conventional extrusion-based 3D printing will provide precise control over the structural and compositional properties of tissue engineering scaffolds [111]. Using a microfluidic-assisted 3D printing approach, a recent study attempted to develop a hydrogel scaffold containing MXene incorporated for skin flap regeneration. By leveraging the temperature-sensitive properties of P(NIPAAm) hydrogels and the photothermal conversion capabilities of MXene nanosheets, the scaffold was said to display near-infrared responsive swelling/shrinkage behavior. According to the research, the scaffold may improve the migration, proliferation, and proangiogenic effects of endothelial cells exposed to near-infrared light by incorporating vascular endothelial growth factor into the hydrogel matrix. According to an in vivo animal study, the scaffold may greatly improve skin flap lifespan by promoting angiogenesis, reducing inflammation, and delaying apoptosis [111].

Interpenetrating networks and other hydrogels that exhibit UCST behavior are less common than LCST systems, but they are typically based on a monomer based on acrylic acid. Similar to LCST systems, the critical temperature of these networks may be altered by adding hydrophilic or hydrophobic co-monomers [104]. An effort was made to 4D-printed hydrogels made of polyacrylic acid that would have both form memory and self-healing capabilities. In order to create a hydrogel, a stereolithography-based resin printer was used to copolymerize the acrylic acid and hexadecyl acrylate (hydrophilic and hydrophobic, respectively) monomers without the need for solvents. This was followed by swelling in water [112]. Additionally, the melting and crystallization temperatures were around human physiological temperature, enabling the shape memory effect of the printed hydrogel. The scientists concluded that by altering the monomers' molar ratio, the shape memory and self-healing capabilities may be triggered at body temperature [112].

5.2.2 pH-Responsive

Ionic pendant groups inside polymeric networks are responsible for the development of pH-responsive hydrogels. After being exposed to an aqueous medium with the proper ionic strength and pH, the pendant groups will ionize and produce preset charges on the network of polymers. This leads to the formation of electrostatic repulsive forces, which cause the hydrogel to inflate and contract in a way that depends on pH [104, 113].

Even small changes in pH may cause a noticeable change in the mesh size of the polymeric networks. The properties of swelling caused by pendant group ionization differ between the two types of pH-responsive hydrogels. In anionic hydrogels, the pendant groups are ionized above the acid dissolution constant (pKa) value and un-ionized below it. Because of the strong osmotic pressure brought on by the availability of the ions, the hydrogel will expand if it is exposed to a medium with a pH greater than its pK a value. In contrast, since the dependent groups in cationic hydrogels are ionized at a pH lower than their pKa value, these hydrogels will swell in environments with a lower pH than the pKa value [113]. The swelling medium's characteristics, as well as the hydrogel's characteristics, such as the ionizable group's charge, concentration, and pKa, will also affect how much an ionic hydrogel swells. The pH and ionic strength of the solution will determine the swelling medium's composition, which is governed by the primary counterions and their valency in solution [113].

The most frequent and prevalent consequences are bacterial infections linked to the use of prosthesis and implanted devices. The key concern with the use of antibiotics is making sure that their release and action are sustained for a long time after surgery. Additionally, prior research has shown that excessive dosages of several antibiotics can be ototoxic and nephrotoxic. Antibiotic systemic exposure in blood and urine is a major safety concern since the loaded doses for the majority of controlled release devices are often high. Therefore, local and targeted delivery of antimicrobial drugs is a more suitable therapy approach [114]. Garcia et al. looked at the creation of stereolithographically 3D-printed hydrogels that are pH-responsive and antibacterial. One of the most popular monomers for introducing pH-responsive behavior is acrylic acid. The scientists claim that the printed hydrogel showed reversible swelling and shrinking in response to pH variations in the environment, and that the degree of swelling or shrinking was immediately connected with the quantity of acrylic acid. The antibacterial properties of the hydrogels were investigated using *Staphylococcus aureus* as a bacterial model [115].

Because of their great volume responsiveness to pH variations in the environment, pH-responsive hydrogels have been thoroughly studied for the regulated administration of medications and treatments. A shift in the surrounding pH, such as that which occurs during passage through the gastrointestinal system, initiates the controlled release of the medication from these loaded hydrogels [104]. In a recent work, Wang et al. [116] suggested creating a hydrogel drug carrier with 3D printing (semi-solid extrusion) and pH-responsive materials. This would allow for site-targeted drug release and allow for the customization of temporal release patterns. The hydrogel's swelling properties were investigated under both intestinal fluids and artificial gastric. Additionally, they asserted that by adjusting the printing process's infill density, controlled release could be accomplished.

5.2.3 Magneto-Responsive

Additionally, the potential for using a magnetic field to both activate and regulate the characteristics of magneto-responsive hydrogels has been investigated. Typically made up of a magnetic component (like magnetite, maghemite, or magnetic iron oxides) and a hydrogel foundation, these types of gels are sensitive to an external magnetic field [117–119]. A magnetic field's accurate and rapid field-induced controllability makes it ideal as an external stimulus [118]. It has been proposed that electromagnetic stimulation is the perfect stimulus, especially for in vivo applications, because it can initiate distant actuation with a fast reaction rate and biocompatibility even at high field strengths [120]. Magnetic particles will begin to build up as soon as an external magnetic field is applied, pushing solvents out of the hydrogel network during compression. Ultimately, the hydrogel's shape will change rapidly. The characteristics of the magneto-responsive hydrogel will depend on the size, composition, concentration, and homogeneity of the magnetic particles inside the hydrogel [105].

Superparamagnetic Fe_3O_4 nanoparticles are one of the most widely utilized magnetic materials for medicinal applications, including tissue engineering [117–119]. However, interparticle magnetic interactions cause superparamagnetic Fe_3O_4 nanoparticles to aggregate, resulting in multicore particles, and magnetic nanoparticle clustering may drastically alter the collective magnetic properties of the particles

[121]. The tendency of these nanoparticles and multicore particles to clump and silt at neutral pH is a major disadvantage [119]. A research suggests that a dynamic tissue scaffold that can replicate the extracellular matrix environment present in the human body might be 3D printed using hydrogels composed of oxidized sodium hyaluronate and glycol chitosan in combination with superparamagnetic iron oxide nanoparticles. The research found that the scaffold's size might change if a magnetic field was applied. Additionally, in vitro investigations showed that the chondrogenic growth of ATDC5 cells encapsulated and cultured inside the hydrogel was significantly influenced by the magnetic field's strength [117].

5.2.4 Moisture or Humidity-Responsive

Hydrogels that are sensitive to moisture or humidity can distort when used as a stimulus in 4D printing, but they will return to their original shape after drying [122]. Acrylic monomers make up the majority of water-responsive hydrogels, also known as superabsorbent polymers, which are weakly crosslinked. They have a very high liquid swelling capacity as a result [120]. Additionally, cellulose and other renewable basic materials have been studied [123].

The swelling ratio of a hydrogel is known to be influenced by the degree of crosslinking; structures with a high degree of crosslinking will have a lower swelling ratio than hydrogels with a lower crosslinking density [124]. For PEG diacrylate hydrogel microstructures created by femtosecond laser direct writing, Lv et al. [125] looked at the potential for humidity-driven mobility. The hydrogel microstructures' voxels have controlled crosslinking density and, consequently, controllable swelling caused by humidity, which may be achieved by varying the manufacturing conditions. They asserted that the hydrogel microstructures' humidity-responsive actuation was stable and reproducible after 10,000 cycles.

Recent research sought to develop a device that would use a hydrogel's swelling capabilities to initiate a self-folding process when exposed to physiological fluids [126]. The authors claim that photopolymerization was used to print a bilayer of a highly swelling sodium acrylate hydrogel and a flexible polyurethane printing resin to create this folding mechanism. Their goal was to create a small-nerve interface cuff electrode, and the soft gel serves as the actuation mechanism, swelling when it comes into contact with fluid following implantation (Fig. 5.5A) [126]. In a similar vein, Yang et al.'s work investigated the hydrogels' capacity to expand as well as the method by which they change form when exposed to an aqueous media. Their goal was to create a collection of fibers rich in cells that resembled skeletal muscle tissue's inherent structure by using gelatin hydrogel printed using 4D technology [127]. The authors claimed that the grooves on the gelatin films were among the most crucial elements in folding. When the films were exposed to water, they began to expand, and after swelling, their thickness altered, but their length and breadth stayed the same. The film's bottom and surface experienced distinct swelling events as a result of the grooved topology. As a result, the gelatin film "rolled up," creating a tubular

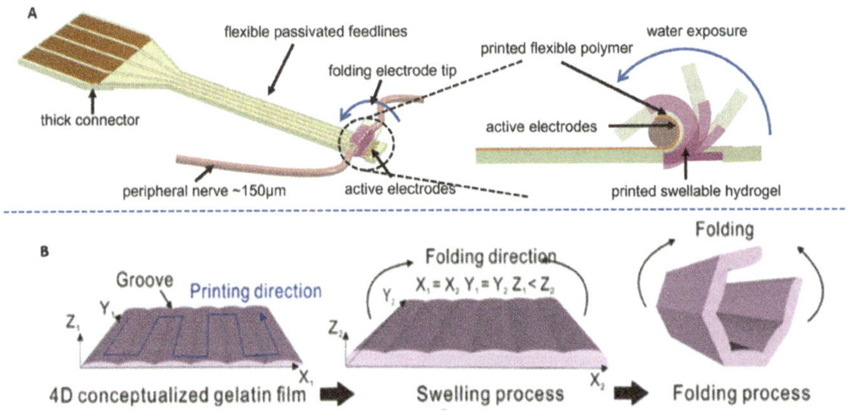

Fig. 5.5 Mechanisms by which moisture-responsive hydrogel changes shape. **A** self-folding electrode that is 4D printed. Reproduced with permission from Ref. [126]. Copyright 2023 Wiley. (Open Access). **B** Gelatin films have a shape-changing mechanism, and their grooves are one of the key elements that cause folding. Reproduced with permission from Ref. [127]. Copyright 2021 Ivyspring International Publisher. (Open Access)

shape (Fig. 5.5B). It should be noted that Li et al. [128] assert that the possibility for humidity or moisture-responsive hydrogels to shift to a single instance would be limited by the continuously high humidity environment employed in in vitro cell growth research and in vivo implantations. More importantly, a certain osmotic pressure is necessary for appropriate cell function [129]. Therefore, restrictions on the amount of humidity or osmotic pressure that may be delivered may limit the degree of shape modification in humidity-responsive hydrogels.

5.3 3D Bioprinting of Hydrogels and Ferrogels

To produce tissue-like structures, cells and biomaterials are simultaneously printed with micrometer accuracy during 3D bioprinting. This method has shown promise in regenerative medicine and tissue engineering for the creation of different transplantable tissues, such as bone, cartilage, and skin. Utilizing 3D bioprinting has several benefits, including precise control over cell dispersion and high-resolution cell deposition. In contrast to the traditional 3D printing techniques previously addressed, bioprinting necessitates a specific technological process that is companionable with the deposition of living cells (Fig. 5.6) [130]. Employing deposition methods such as extrusion, laser-induced forward transfer (LIFT), and inkjet-based printing, the biofabrication of hydrogels as a vehicle for cells and/or bioactive substances has been studied. Because of the rheology and post-curing rate of the hydrogel-based bioinks that are used, each of these methods necessitates highly precise preconditions. This is to ensure that 3D constructions are manufactured with reliability [131]. Notwithstanding their application in bioprinting, each methodology has drawbacks of its own, and no one technology enables the biofabrication of synthetic tissues at all sizes and complexity [131, 132]. For instance, compared to other printer types, only a small number of bioprinters (that use lasers) are accessible because of their enhanced price and complexity [80]. Similarly, because of their tiny orifice width, which prevents them from printing cells at high densities, inkjet printers have the potential to inflict substantial cell mutilation and demise, along with cell sedimentation and aggregation [131, 132]. Likewise, extrusion-based methods may result in greater pressure reductions at the nozzle or cause cell deformation from shear stress, both of which might be detrimental to embedded cells. The main factors to be taken into account when selecting the right cells for bioprinting are the degree to which the bioprinted cells can grow or maintain their functions in vivo under ideal microenvironments and the degree to which these printed cells can replicate the physiological state of cells in vivo. To prevent any unfavorable immunological reactions, bioprinting should ideally be done using the patient's own cells. Since not all cell types can regenerate after injury, the most promising source of cells is stem cells. (Fig. 5.6) [130].

The right bioinks are necessary for biofabrication and cell printing because they must enable the culture medium and embedded cells to properly diffuse nutrients, oxygen, and metabolic wastes while shielding the implanted cells from outward mechanical stresses during the printing process [133]. In addition to creating a three-dimensional structure that is mechanically robust and able to maintain structural integrity, the bioink must also be non-cytotoxic, biocompatible, and allow the attachment and proliferation of the loaded cells. Characterizing and developing new bioinks has received increased attention recently due to the lack of materials suitable

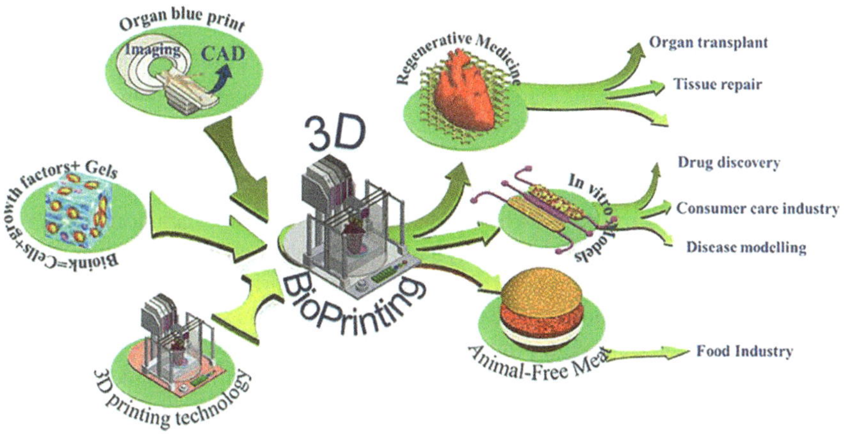

Fig. 5.6 3D bioprinting combines conventional 3D printing, imaging, and cell-gel to produce efficient tissue for regenerative treatment, pharmaceutical preclinical drug screening, and animal-free meat. Reproduced with permission from Ref. [130]. Copyright 2021 Frontiers. (Open Access)

for bioprinting; this lack was believed to be a major barrier preventing the advancement of cell printing and biofabrication technologies [133, 134]. The term "bioink," which was originally used in 2003 in relation to organ printing, was first introduced by Mironov [135] and then subsequently reviewed by Groll et al. [134]. The initial plan was to use a technology called "bioprinting" to place tissue spheroids or live cells as the "bioink" after printing a hydrogel.

Therefore, the cellular component that was placed on or inside hydrogels in three dimensions was originally called bioink. More significantly, during formulation and processing, biomaterials that may be regarded as bioinks must serve as a vehicle for cell distribution. Groll et al. suggested that (bio)materials that can be printed and then seeded with cells after printing but are not made directly with cells should be categorized as biomaterial inks rather than bioinks in order to provide a clearer difference [134]. To create an inherent mechanical support inside the build, these biomaterial inks might be used in hybrid techniques in combination with bioink production [134]. "A formulation of cells appropriate for processing through an automated biofabrication technology that may also contain biologically active components and biomaterials" is another definition they suggested be used for bioinks. The hydrogel-based bioink's crosslinking capabilities, gelation, and viscosity are some of the characteristics that must be taken into account before printing, since they can significantly affect the viability, morphology of cells and proliferation after printing. Making a hydrogel that can support and shield cells while also acting as a structurally sound scaffold is difficult. This is due to the fact that various tasks are linked to distinct mechanical needs [131, 136].

5.4 Challenges and Future Prospects

The advancement of additive manufacturing technologies, particularly 3D printing, 4D printing, and bioprinting, has enabled the fabrication of smart materials such as hydrogels and ferrogels with applications in drug delivery, soft robotics and tissue engineering. However, these technologies face notable challenges and limitations. In 3D printing of hydrogels and ferrogels, a primary concern lies in the rheological properties of the materials, which must strike a balance between printability and structural fidelity. Hydrogels, by nature, have high water content and low mechanical strength, which often leads to deformation or collapse during or after printing [137]. Ferrogels introduce further complexity due to the inclusion of magnetic nanoparticles, which can interfere with print resolution and homogeneity. The advent of 4D printing, which adds a time-dependent, responsive behavior to the printed structure, introduces additional difficulties in predicting and controlling shape transformation, especially when exposed to multiple stimuli. These transformations depend heavily on material anisotropy and environmental sensitivity, making the fabrication process less reliable and reproducible [92]. Bioprinting presents its own set of challenges, particularly in cell viability, bioink formulation, and maintaining sterility. The integration of living cells into hydrogel matrices demands optimal conditions that preserve cellular function, while also ensuring the mechanical stability of the construct during and after printing [138]. Furthermore, long-term biocompatibility, vascularization, and immune responses remain critical barriers to clinical application. Altogether, while 3D, 4D, and bioprinting technologies offer potential, the limitations in material behavior, structural integrity, process control, and biological functionality must be systematically addressed to enable widespread and reliable implementation.

5.5 Summary of the Chapter

The production of hydrogel network by 3D printing has grown in popularity among biomedical researchers due to its ability to produce detailed, complicated, and extremely adjustable scaffolds for tissue engineering that may enable adhesion and encourage cell infiltration. The concept of 4D printing was first introduced in 2014. The technique may be characterized as a production method that blends the usage of materials that respond to stimuli, additive manufacturing, and mathematical modeling. Several production methods are now being researched and developed for 4D printing; these methods are essentially based on the same additive manufacturing processes that are employed in 3D printing. When compared to 3D printing, the main benefit of 4D printing is its capacity to produce dynamic structures. It is commonly known that several elements will affect the 4D printing method, and the stimulus-material interaction is one of the most important ones. This is due to the fact that the 4D was introduced into the printing progression by the responsive material. Unlike regular 3D printing, bioprinting will need a new technological method that is compatible with the deposition of live cells since its purpose is to produce tissue-like structures employing microscopic entities of biomaterials and cells distributed with micrometer accuracy. 3D bioprinting's use has several benefits, including precise control over cell distribution and high-resolution cell deposition. All synthetic tissue sizes and complexities cannot be biofabricated using a single bioprinting technology, and each approach has drawbacks of its own. For biofabrication and cell printing, selecting the right cells and bioinks is essential because they must allow for the proper diffusion of oxygen, nutrients, and metabolic waste products between the culture medium and embedded cells while simultaneously protecting the implanted cells, subjected to external mechanical loads during the process of printing. Bioprinting has revealed considerable promise in tissue engineering and regenerative medicine, and observations from limited in vivo experiments as well as in vitro investigations showed that artificial organ transplantation and bioprinting may be possible in the future.

References

1. A. Gajakosh, S. Erannagari, R.S. Kumar, Thyagaraj N.R., Mallaradhya H.M., Rudresha S., *An Overview of 3D-Printed Smart Polymers and Composites* (2023), pp. 130–148. https://doi.org/10.4018/978-1-6684-6009-2.ch008
2. I. Tasneem, A. Ariz, D. Bharti, A. Haleem, M. Javaid, S. Bahl, 3D printing technology and its significant applications in the context of healthcare education. J. Industr. Integr. Manag. **08**, 113–130 (2023). https://doi.org/10.1142/S2424862221500159
3. T.-C. Ho, C.-C. Chang, H.-P. Chan, T.-W. Chung, C.-W. Shu, K.-P. Chuang, T.-H. Duh, M.-H. Yang, Y.-C. Tyan, Hydrogels: properties and applications in biomedicine. Molecules **27**, 2902 (2022). https://doi.org/10.3390/molecules27092902
4. S. Awasthi, A review on hydrogels and Ferrogels for biomedical applications. JOM **73**, 2440–2451 (2021). https://doi.org/10.1007/s11837-021-04734-7
5. D. Yang, Recent advances in hydrogels. Chem. Mater. **34**, 1987–1989 (2022). https://doi.org/10.1021/acs.chemmater.2c00188
6. A. Agrawal, C.M. Hussain, Materials and technologies for flexible and wearable sensors, in *Flexible and Wearable Sensors*, (CRC Press, Boca Raton, 2023), pp. 21–34. https://doi.org/10.1201/9781003299455-2
7. J.K. Placone, A.J. Engler, Recent advances in extrusion-based 3D printing for biomedical applications. Adv. Healthc. Mater. **7** (2018). https://doi.org/10.1002/adhm.201701161
8. Z. Wang, X. Jin, R. Dai, J.F. Holzman, K. Kim, An ultrafast hydrogel photocrosslinking method for direct laser bioprinting. RSC Adv. **6**, 21099–21104 (2016). https://doi.org/10.1039/C5RA24910D
9. X. Mo, L. Ouyang, Z. Xiong, T. Zhang, Advances in digital light processing of hydrogels. Biomed. Mater. **17**, 042002 (2022). https://doi.org/10.1088/1748-605X/ac6b04
10. S. Naghieh, M. Sarker, N.K. Sharma, Z. Barhoumi, X. Chen, Printability of 3D printed hydrogel scaffolds: Influence of hydrogel composition and printing parameters. Appl. Sci. **10**, 292 (2019). https://doi.org/10.3390/app10010292
11. B. Pavan Kalyan, L. Kumar, 3D printing: applications in tissue engineering, medical devices, and drug delivery. AAPS Pharm. Sci. Tech. **23**, 92 (2022). https://doi.org/10.1208/s12249-022-02242-8
12. O. Kapusta, A. Jarosz, K. Stadnik, D.A. Giannakoudakis, B. Barczyński, M. Barczak, Antimicrobial natural hydrogels in biomedicine: properties, applications, and challenges—a concise review. Int. J. Mol. Sci. **24**, 2191 (2023). https://doi.org/10.3390/ijms24032191
13. G. Kaliaraj, D. Shanmugam, A. Dasan, K. Mosas, Hydrogels—a promising materials for 3D printing technology. Gels **9**, 260 (2023). https://doi.org/10.3390/gels9030260
14. X.N. Zhang, Q. Zheng, Z.L. Wu, Recent advances in 3D printing of tough hydrogels: a review. Compos. Part B Eng. **238**, 109895 (2022). https://doi.org/10.1016/j.compositesb.2022.109895
15. A. Agrawal, C.M. Hussain, 3D-printed hydrogel for diverse applications: a review. Gels **9**, 960 (2023). https://doi.org/10.3390/gels9120960
16. S. Awasthi, In vitro and in vivo therapeutics of double-layered hydrogels. Adv. Ther. (Weinh) **8** (2025). https://doi.org/10.1002/adtp.202400266
17. M.L. Bedell, A.L. Torres, K.J. Hogan, Z. Wang, B. Wang, A.J. Melchiorri, K.J. Grande-Allen, A.G. Mikos, Human gelatin-based composite hydrogels for osteochondral tissue engineering and their adaptation into bioinks for extrusion, inkjet, and digital light processing bioprinting. Biofabrication **14**, 045012 (2022). https://doi.org/10.1088/1758-5090/ac8768
18. T. Pose-Boirazian, J. Martínez-Costas, G. Eibes, 3D printing: an emerging technology for biocatalyst immobilization. Macromol. Biosci. **22** (2022). https://doi.org/10.1002/mabi.202200110
19. Y. Cheng, X. Shi, X. Jiang, X. Wang, H. Qin, Printability of a cellulose derivative for extrusion-based 3D printing: the application on a biodegradable support material. Front. Mater. **7** (2020). https://doi.org/10.3389/fmats.2020.00086
20. A.H. Choi, B. Ben-Nissan, *3D, 4D Printing, and Bioprinting of Hydrogels* (2024), pp. 29–59. https://doi.org/10.1007/978-981-97-1730-9_2

21. A.H. Choi, B. Ben-Nissan, *Hydrogel for Biomedical Applications: 3D/4D Printing, Self-Healing, Microrobots, and Nanogenerators*, 1st edn. (Springer Nature Singapore, Singapore, 2024). https://doi.org/10.1007/978-981-97-1730-9
22. A.P. Dhand, M.D. Davidson, J.A. Burdick, Lithography-based 3D printing of hydrogels. Nat. Rev. Bioeng. (2025). https://doi.org/10.1038/s44222-024-00251-9
23. T. Billiet, M. Vandenhaute, J. Schelfhout, S. Van Vlierberghe, P. Dubruel, A review of trends and limitations in hydrogel-rapid prototyping for tissue engineering. Biomaterials **33**, 6020–6041 (2012). https://doi.org/10.1016/j.biomaterials.2012.04.050
24. W.-Y. Yeong, C.-K. Chua, K.-F. Leong, M. Chandrasekaran, Rapid prototyping in tissue engineering: challenges and potential. Trends Biotechnol. **22**, 643–652 (2004). https://doi.org/10.1016/j.tibtech.2004.10.004
25. R. Landers, U. Hübner, R. Schmelzeisen, R. Mülhaupt, Rapid prototyping of scaffolds derived from thermoreversible hydrogels and tailored for applications in tissue engineering. Biomaterials **23**, 4437–4447 (2002). https://doi.org/10.1016/S0142-9612(02)00139-4
26. S.M. Peltola, F.P.W. Melchels, D.W. Grijpma, M. Kellomäki, A review of rapid prototyping techniques for tissue engineering purposes. Ann. Med. **40**, 268–280 (2008). https://doi.org/10.1080/07853890701881788
27. I.-H. Liu, S.-H. Chang, H.-Y. Lin, Chitosan-based hydrogel tissue scaffolds made by 3D plotting promotes osteoblast proliferation and mineralization. Biomed. Mater. **10**, 035004 (2015). https://doi.org/10.1088/1748-6041/10/3/035004
28. T. Pan, W. Song, H. Xin, H. Yu, H. Wang, D. Ma, X. Cao, Y. Wang, MicroRNA-activated hydrogel scaffold generated by 3D printing accelerates bone regeneration. Bioact. Mater. **10**, 1–14 (2022). https://doi.org/10.1016/j.bioactmat.2021.08.034
29. X. Xuan, H. Zhang, X. Xu, Z. Pan, Y. Luo, Y. Li, L. Sun, Tuning polyacrylamide precursor viscoelasticity using Nanoclay for extrusion-based 3D printing of hydrogels. J. Polym. Sci. **63**, 1924–1937 (2025). https://doi.org/10.1002/pol.20250147
30. R.C. Steiner, J.T. Buchen, E.R. Phillips, C.R. Fellin, X. Yuan, S.H. Jariwala, FRESH extrusion 3D printing of type-1 collagen hydrogels photocrosslinked using ruthenium. PLoS One **20**, e0317350 (2025). https://doi.org/10.1371/journal.pone.0317350
31. P. Panyamao, S. Charumanee, J. Ruangsuriya, C. Saenjum, Coffee parchment-derived nanocellulose as reinforcing agent in hydrogel inks for extrusion-based 3D printing of biphasic osteochondral scaffolds. Carbohydr. Polym. Technol. Appl. **10**, 100752 (2025). https://doi.org/10.1016/j.carpta.2025.100752
32. P. Calvert, Inkjet printing for materials and devices. Chem. Mater. **13**, 3299–3305 (2001). https://doi.org/10.1021/cm0101632
33. T.-S. Jang, H.-D. Jung, H.M. Pan, W.T. Han, S. Chen, J. Song, 3D printing of hydrogel composite systems: recent advances in technology for tissue engineering. Int. J. Bioprint. **4**, 126 (2024). https://doi.org/10.18063/ijb.v4i1.126
34. J. Li, M. Chen, X. Fan, H. Zhou, Recent advances in bioprinting techniques: approaches, applications and future prospects. J. Transl. Med. **14**, 271 (2016). https://doi.org/10.1186/s12967-016-1028-0
35. M. Nakamura, A. Kobayashi, F. Takagi, A. Watanabe, Y. Hiruma, K. Ohuchi, Y. Iwasaki, M. Horie, I. Morita, S. Takatani, Biocompatible inkjet printing technique for designed seeding of individual living cells. Tissue Eng. **11**, 1658–1666 (2005). https://doi.org/10.1089/ten.2005.11.1658
36. R.E. Saunders, J.E. Gough, B. Derby, Delivery of human fibroblast cells by piezoelectric drop-on-demand inkjet printing. Biomaterials **29**, 193–203 (2008). https://doi.org/10.1016/j.biomaterials.2007.09.032
37. X. Wang, Q. Ao, X. Tian, J. Fan, Y. Wei, W. Hou, H. Tong, S. Bai, 3D bioprinting technologies for hard tissue and organ engineering. Materials **9**, 802 (2016). https://doi.org/10.3390/ma9100802
38. A. Negro, T. Cherbuin, M.P. Lutolf, 3D inkjet printing of complex, cell-laden hydrogel structures. Sci. Rep. **8**, 17099 (2018). https://doi.org/10.1038/s41598-018-35504-2

39. P.J. Smith, A. Morrin, Reactive inkjet printing. J. Mater. Chem. **22**, 10965 (2012). https://doi.org/10.1039/c2jm30649b
40. G.L. Duffy, H. Liang, R.L. Williams, D.A. Wellings, K. Black, 3D reactive inkjet printing of poly-ε-lysine/gellan gum hydrogels for potential corneal constructs. Mater. Sci. Eng. C **131**, 112476 (2021). https://doi.org/10.1016/j.msec.2021.112476
41. M.Y. Teo, L. Stuart, K.C. Aw, J. Stringer, Micro-reactive inkjet printing of three-dimensional hydrogel structures. MRS Adv. **3**, 1575–1581 (2018). https://doi.org/10.1557/adv.2017.628
42. Y. Lee, J.A. Park, T. Tuladhar, S. Jung, Sonochemical degradation of Gelatin Methacryloyl to control viscoelasticity for inkjet bioprinting. Macromol. Biosci. **23** (2023). https://doi.org/10.1002/mabi.202200509
43. F. Cai, S. Heid, A.R. Boccaccini, Potential of Laponite® incorporated oxidized alginate–gelatin (ADA-GEL) composite hydrogels for extrusion-based 3D printing. J. Biomed. Mater. Res. B Appl. Biomater. **109**, 1090–1104 (2021). https://doi.org/10.1002/jbm.b.34771
44. S. Zhou, C. Han, Z. Ni, C. Yang, Y. Ni, Y. Lv, Gelatin-oxidized Nanocellulose hydrogels suitable for extrusion-based 3D bioprinting. PRO **10**, 2216 (2022). https://doi.org/10.3390/pr10112216
45. L. Dong, Z. Bu, Y. Xiong, H. Zhang, J. Fang, H. Hu, Z. Liu, X. Li, Facile extrusion 3D printing of gelatine methacrylate/Laponite nanocomposite hydrogel with high concentration nanoclay for bone tissue regeneration. Int. J. Biol. Macromol. **188**, 72–81 (2021). https://doi.org/10.1016/j.ijbiomac.2021.07.199
46. G. Falcone, P. Mazzei, A. Piccolo, T. Esposito, T. Mencherini, R.P. Aquino, P. Del Gaudio, P. Russo, Advanced printable hydrogels from pre-crosslinked alginate as a new tool in semi solid extrusion 3D printing process. Carbohydr. Polym. **276**, 118746 (2022). https://doi.org/10.1016/j.carbpol.2021.118746
47. R.D. Murphy, S. Kimmins, A.J. Hibbitts, A. Heise, 3D-extrusion printing of stable constructs composed of photoresponsive polypeptide hydrogels. Polym. Chem. **10**, 4675–4682 (2019). https://doi.org/10.1039/C9PY00796B
48. X. Hu, Z. Yang, S. Kang, M. Jiang, Z. Zhou, J. Gou, D. Hui, J. He, Cellulose hydrogel skeleton by extrusion 3D printing of solution. Nanotechnol. Rev. **9**, 345–353 (2020). https://doi.org/10.1515/ntrev-2020-0025
49. X. Cui, J. Li, Y. Hartanto, M. Durham, J. Tang, H. Zhang, G. Hooper, K. Lim, T. Woodfield, Advances in extrusion 3D bioprinting: a focus on multicomponent hydrogel-based bioinks. Adv. Healthc. Mater. **9** (2020). https://doi.org/10.1002/adhm.201901648
50. C. Tetyczka, K. Brisberger, M. Reiser, M. Zettl, R. Jeitler, C. Winter, D. Kolb, G. Leitinger, M. Spoerk, E. Roblegg, Itraconazole nanocrystals on hydrogel contact lenses via inkjet printing: Implications for ophthalmic drug delivery. ACS Appl. Nano Mater. **5**, 9435–9446 (2022). https://doi.org/10.1021/acsanm.2c01715
51. R. Suntornnond, W.L. Ng, X. Huang, C.H.E. Yeow, W.Y. Yeong, Improving printability of hydrogel-based bio-inks for thermal inkjet bioprinting applications *via* saponification and heat treatment processes. J. Mater. Chem. B **10**, 5989–6000 (2022). https://doi.org/10.1039/D2TB00442A
52. F. Chen, S. Gu, Q. Zhang, T. Liu, Z. Liu, T. Kuang, A comparison study of hyaluronic acid hydrogel exquisite micropatterns with photolithography and light-cured inkjet printing methods. E-Polymers **22**, 332–341 (2022). https://doi.org/10.1515/epoly-2022-0027
53. M.Y. Teo, S. Kee, N. RaviChandran, L. Stuart, K.C. Aw, J. Stringer, Enabling free-standing 3D hydrogel microstructures with microreactive inkjet printing. ACS Appl. Mater. Interfaces **12**, 1832–1839 (2020). https://doi.org/10.1021/acsami.9b17192
54. Y. Nakagawa, S. Ohta, M. Nakamura, T. Ito, 3D inkjet printing of star block copolymer hydrogels cross-linked using various metallic ions. RSC Adv. **7**, 55571–55576 (2017). https://doi.org/10.1039/C7RA11509A
55. T. Jiao, Q. Lian, T. Zhao, H. Wang, D. Li, Preparation, mechanical and biological properties of inkjet printed alginate/Gelatin hydrogel. J. Bionic. Eng. **18**, 574–583 (2021). https://doi.org/10.1007/s42235-021-0036-9

56. S. Yoon, J.A. Park, H. Lee, W.H. Yoon, D.S. Hwang, S. Jung, Inkjet–spray hybrid printing for 3D freeform fabrication of multilayered hydrogel structures. Adv. Healthc. Mater. **7** (2018). https://doi.org/10.1002/adhm.201800050
57. X. Peng, T. Liu, Q. Zhang, C. Shang, Q. Bai, H. Wang, Surface patterning of hydrogels for programmable and complex shape deformations by ion inkjet printing. Adv. Funct. Mater. **27** (2017). https://doi.org/10.1002/adfm.201701962
58. L.M. Kalossaka, A.A. Mohammed, G. Sena, L. Barter, C. Myant, 3D printing nanocomposite hydrogels with lattice vascular networks using stereolithography. J. Mater. Res. **36**, 4249–4261 (2021). https://doi.org/10.1557/s43578-021-00411-2
59. I. Karakurt, A. Aydoğdu, S. Çıkrıkcı, J. Orozco, L. Lin, Stereolithography (SLA) 3D printing of ascorbic acid loaded hydrogels: a controlled release study. Int. J. Pharm. **584**, 119428 (2020). https://doi.org/10.1016/j.ijpharm.2020.119428
60. G. Burke, D.M. Devine, I. Major, Effect of Stereolithography 3D printing on the properties of PEGDMA hydrogels. Polymers (Basel) **12**, 2015 (2020). https://doi.org/10.3390/polym12092015
61. L.S.S.M. Magalhães, F.E.P. Santos, C.d.M.V. Elias, S. Afewerki, G.F. Sousa, A.S.A. Furtado, F.R. Marciano, A.O. Lobo, Printing 3D hydrogel structures employing low-cost Stereolithography technology. J. Funct. Biomater. **11**, 12 (2020). https://doi.org/10.3390/jfb11010012
62. A.S. Alketbi, Y. Shi, H. Li, A. Raza, T. Zhang, Impact of PEGDA photopolymerization in micro-stereolithography on 3D printed hydrogel structure and swelling. Soft. Matter. **17**, 7188–7195 (2021). https://doi.org/10.1039/D1SM00483B
63. Z. Sun, Y. Lu, Q. Zhao, J. Wu, A new stereolithographic 3D printing strategy for hydrogels with a large mechanical tunability and self-weldability. Addit. Manuf. **50**, 102563 (2022). https://doi.org/10.1016/j.addma.2021.102563
64. N. Risangud, T. Lertwimol, S. Sitthisang, W. Wongvitvichot, P. Uppanan, S. Tanodekaew, The preparation of 3D-printed self-healing hydrogels composed of carboxymethyl chitosan and oxidized dextran via stereolithography for biomedical applications. Int. J. Biol. Macromol. **292**, 139251 (2025). https://doi.org/10.1016/j.ijbiomac.2024.139251
65. X. Yue, M. Liu, Y. Xia, Z. Xiong, Influences of two sizes of nano-Fe_3O_4 on stereolithography 3D printing properties of gel photoresist for photothermal-responsive hydrogel. Mater. Today Commun. **39**, 108881 (2024). https://doi.org/10.1016/j.mtcomm.2024.108881
66. M. Caprioli, I. Roppolo, A. Chiappone, L. Larush, C.F. Pirri, S. Magdassi, 3D-printed self-healing hydrogels via digital light processing. Nat. Commun. **12**, 2462 (2021). https://doi.org/10.1038/s41467-021-22802-z
67. C. Mendes-Felipe, J. Oliveira, I. Etxebarria, J.L. Vilas-Vilela, S. Lanceros-Mendez, State-of-the-art and future challenges of UV curable polymer-based smart materials for printing technologies. Adv. Mater. Technol. **4** (2019). https://doi.org/10.1002/admt.201800618
68. H.H. Hwang, W. Zhu, G. Victorine, N. Lawrence, S. Chen, 3D-printing of functional biomedical microdevices via light- and extrusion-based approaches. Small Methods **2** (2018). https://doi.org/10.1002/smtd.201700277
69. Y. Lu, G. Mapili, G. Suhali, S. Chen, K. Roy, A digital micro-mirror device-based system for the microfabrication of complex, spatially patterned tissue engineering scaffolds. J. Biomed. Mater. Res. A **77A**, 396–405 (2006). https://doi.org/10.1002/jbm.a.30601
70. F.P.W. Melchels, J. Feijen, D.W. Grijpma, A review on stereolithography and its applications in biomedical engineering. Biomaterials **31**, 6121–6130 (2010). https://doi.org/10.1016/j.biomaterials.2010.04.050
71. M.T. Sultan, O.J. Lee, J.S. Lee, C.H. Park, Three-dimensional digital light-processing bioprinting using silk fibroin-based bio-ink: recent advancements in biomedical applications. Biomedicine **10**, 3224 (2022). https://doi.org/10.3390/biomedicines10123224
72. S.H. Kim, Y.K. Yeon, J.M. Lee, J.R. Chao, Y.J. Lee, Y.B. Seo, M.T. Sultan, O.J. Lee, J.S. Lee, S. Yoon, I.-S. Hong, G. Khang, S.J. Lee, J.J. Yoo, C.H. Park, Precisely printable and biocompatible silk fibroin bioink for digital light processing 3D printing. Nat. Commun. **9**, 1620 (2018). https://doi.org/10.1038/s41467-018-03759-y

73. H. Hong, Y.B. Seo, D.Y. Kim, J.S. Lee, Y.J. Lee, H. Lee, O. Ajiteru, M.T. Sultan, O.J. Lee, S.H. Kim, C.H. Park, Digital light processing 3D printed silk fibroin hydrogel for cartilage tissue engineering. Biomaterials **232**, 119679 (2020). https://doi.org/10.1016/j.biomaterials.2019.119679
74. J. Koffler, W. Zhu, X. Qu, O. Platoshyn, J.N. Dulin, J. Brock, L. Graham, P. Lu, J. Sakamoto, M. Marsala, S. Chen, M.H. Tuszynski, Biomimetic 3D-printed scaffolds for spinal cord injury repair. Nat. Med. **25**, 263–269 (2019). https://doi.org/10.1038/s41591-018-0296-z
75. D. Xue, J. Zhang, Y. Wang, D. Mei, Digital light processing-based 3D printing of cell-seeding hydrogel scaffolds with regionally varied stiffness. ACS Biomater. Sci. Eng. **5**, 4825–4833 (2019). https://doi.org/10.1021/acsbiomaterials.9b00696
76. S. Vaupel, R. Mau, S. Kara, H. Seitz, U. Kragl, J. Meyer, 3D printed and stimulus responsive drug delivery systems based on synthetic polyelectrolyte hydrogels manufactured *via* digital light processing. J. Mater. Chem. B **11**, 6547–6559 (2023). https://doi.org/10.1039/D3TB00285C
77. K.Y. Choi, O. Ajiteru, H. Hong, Y.J. Suh, M.T. Sultan, H. Lee, J.S. Lee, Y.J. Lee, O.J. Lee, S.H. Kim, C.H. Park, A digital light processing 3D-printed artificial skin model and full-thickness wound models using silk fibroin bioink. Acta Biomater. **164**, 159–174 (2023). https://doi.org/10.1016/j.actbio.2023.04.034
78. A. Bhusal, E. Dogan, H.-A. Nguyen, O. Labutina, D. Nieto, A. Khademhosseini, A.K. Miri, Multi-material digital light processing bioprinting of hydrogel-based microfluidic chips. Biofabrication **14**, 014103 (2022). https://doi.org/10.1088/1758-5090/ac2d78
79. X. Zhang, W. Wu, Y. Huang, X. Yang, M. Gou, Antheraea pernyi silk fibroin bioinks for digital light processing 3D printing. Int. J. Bioprint. **9**, 760 (2023). https://doi.org/10.18063/ijb.760
80. Z. Xiang, N. Li, Y. Rong, L. Zhu, X. Huang, 3D-printed high-toughness double network hydrogels via digital light processing. Colloids Surf. A Physicochem. Eng. Asp. **639**, 128329 (2022). https://doi.org/10.1016/j.colsurfa.2022.128329
81. N. Lopez-Larrea, M. Criado-Gonzalez, A. Dominguez-Alfaro, N. Alegret, I. del Agua, B. Marchiori, D. Mecerreyes, Digital light 3D printing of PEDOT-based photopolymerizable inks for biosensing. ACS Appl. Polym. Mater. **4**, 6749–6759 (2022). https://doi.org/10.1021/acsapm.2c01170
82. K. Sekmen, T. Rehbein, M. Johlitz, A. Lion, A. Constantinescu, Thermal analysis and shrinkage characterization of the photopolymers for DLP additive manufacturing processes. Contin. Mech. Thermodyn. **36**, 351–368 (2024). https://doi.org/10.1007/s00161-022-01137-0
83. A. Tang, J. Li, J. Li, S. Zhao, W. Liu, T. Liu, J. Wang, Y. Liu, Nanocellulose/PEGDA aerogel scaffolds with tunable modulus prepared by stereolithography for three-dimensional cell culture. J. Biomater. Sci. Polym. Ed. **30**, 797–814 (2019). https://doi.org/10.1080/09205063.2019.1602904
84. Z. Sun, Q. Zhao, S. Ma, J. Wu, DLP 3D printed hydrogels with hierarchical structures post-programmed by lyophilization and ionic locking. Mater. Horiz. **10**, 179–186 (2023). https://doi.org/10.1039/D2MH00962E
85. M. Dong, Y. Han, X.P. Hao, H.C. Yu, J. Yin, M. Du, Q. Zheng, Z.L. Wu, Digital light processing 3D printing of tough supramolecular hydrogels with sophisticated architectures as impact-absorption elements. Adv. Mater. **34** (2022). https://doi.org/10.1002/adma.202204333
86. Z. Guo, C. Ma, W. Xie, A. Tang, W. Liu, An effective DLP 3D printing strategy of high strength and toughness cellulose hydrogel towards strain sensing. Carbohydr. Polym. **315**, 121006 (2023). https://doi.org/10.1016/j.carbpol.2023.121006
87. D. Cafiso, A.A. Septevani, C. Noè, T. Schiller, C.F. Pirri, I. Roppolo, A. Chiappone, 3D printing of fully cellulose-based hydrogels by digital light processing. Sustain. Mater. Technol. **32**, e00444 (2022). https://doi.org/10.1016/j.susmat.2022.e00444
88. L. Wang, Q. Wang, A. Slita, O. Backman, Z. Gounani, E. Rosqvist, J. Peltonen, S. Willför, C. Xu, J.M. Rosenholm, X. Wang, Digital light processing (DLP) 3D-fabricated antimicrobial hydrogel with a sustainable resin of methacrylated woody polysaccharides and hybrid silver-lignin nanospheres. Green Chem. **24**, 2129–2145 (2022). https://doi.org/10.1039/D1GC03841A

89. P. Fu, H. Li, J. Gong, Z. Fan, A.T. Smith, K. Shen, T.O. Khalfalla, H. Huang, X. Qian, J.R. McCutcheon, L. Sun, 4D printing of polymers: techniques, materials, and prospects. Prog. Polym. Sci. **126**, 101506 (2022). https://doi.org/10.1016/j.progpolymsci.2022.101506
90. R. Sajjad, S.T. Chauhdary, M.T. Anwar, A. Zahid, A.A. Khosa, M. Imran, M.H. Sajjad, A review of 4D printing—technologies, shape shifting, smart polymer based materials, and biomedical applications. Adv. Industr. Eng. Polymer Res. **7**, 20–36 (2024). https://doi.org/10.1016/j.aiepr.2023.08.002
91. Y.S. Lui, W.T. Sow, L.P. Tan, Y. Wu, Y. Lai, H. Li, 4D printing and stimuli-responsive materials in biomedical aspects. Acta Biomater. **92**, 19–36 (2019). https://doi.org/10.1016/j.actbio.2019.05.005
92. F. Momeni, S.M. Mehdi Hassani.N, X. Liu, J. Ni, A review of 4D printing. Mater. Des. **122**, 42–79 (2017). https://doi.org/10.1016/j.matdes.2017.02.068
93. A. Khademhosseini, *3D Bioprinting in Regenerative Engineering* (CRC Press, Boca Raton: Taylor & Francis, 2018). https://doi.org/10.1201/b21916
94. S. Tibbits, 4D printing: multi-material shape change. Archit. Des. **84**, 116–121 (2014). https://doi.org/10.1002/ad.1710
95. G. Liu, Y. Zhao, G. Wu, J. Lu, Origami and 4D printing of elastomer-derived ceramic structures. Sci. Adv. **4** (2018). https://doi.org/10.1126/sciadv.aat0641
96. J.P. Vacanti, R. Langer, Tissue engineering: the design and fabrication of living replacement devices for surgical reconstruction and transplantation. Lancet **354**, S32–S34 (1999). https://doi.org/10.1016/S0140-6736(99)90247-7
97. S. Awasthi, Unveiling the development principles and mechanistic understanding of controlled drug delivery strategies for chronic bone defects and diabetic wound management. Mater. Adv. **6**, 5831–5856 (2025). https://doi.org/10.1039/D5MA00490J
98. H. Thérien-Aubin, Z.L. Wu, Z. Nie, E. Kumacheva, Multiple shape transformations of composite hydrogel sheets. J. Am. Chem. Soc. **135**, 4834–4839 (2013). https://doi.org/10.1021/ja400518c
99. Z.U. Arif, M.Y. Khalid, A. Zolfagharian, M. Bodaghi, 4D bioprinting of smart polymers for biomedical applications: recent progress, challenges, and future perspectives. React. Funct. Polym. **179**, 105374 (2022). https://doi.org/10.1016/j.reactfunctpolym.2022.105374
100. Y. Lu, A.A. Aimetti, R. Langer, Z. Gu, Bioresponsive materials. Nat. Rev. Mater. **2**, 16075 (2016). https://doi.org/10.1038/natrevmats.2016.75
101. B. Mirani, E. Pagan, B. Currie, M.A. Siddiqui, R. Hosseinzadeh, P. Mostafalu, Y.S. Zhang, A. Ghahary, M. Akbari, An advanced multifunctional hydrogel-based dressing for wound monitoring and drug delivery. Adv. Healthc. Mater. **6** (2017). https://doi.org/10.1002/adhm.201700718
102. N. Contessi Negrini, L. Bonetti, L. Contili, S. Farè, 3D printing of methylcellulose-based hydrogels. Bioprinting **10**, e00024 (2018). https://doi.org/10.1016/j.bprint.2018.e00024
103. N.A. Peppas, *Hydrogels in Medicine and Pharmacy* (CRC Press, 2019). https://doi.org/10.1201/9780429285097
104. M.C. Koetting, J.T. Peters, S.D. Steichen, N.A. Peppas, Stimulus-responsive hydrogels: theory, modern advances, and applications. Mater. Sci. Eng. R: Rep. **93**, 1–49 (2015). https://doi.org/10.1016/j.mser.2015.04.001
105. S. Chang, S. Wang, Z. Liu, X. Wang, Advances of stimulus-responsive hydrogels for bone defects repair in tissue engineering. Gels **8**, 389 (2022). https://doi.org/10.3390/gels8060389
106. A.K. Bajpai, S.K. Shukla, S. Bhanu, S. Kankane, Responsive polymers in controlled drug delivery. Prog. Polym. Sci. **33**, 1088–1118 (2008). https://doi.org/10.1016/j.progpolymsci.2008.07.005
107. F.-J. Xu, E.-T. Kang, K.-G. Neoh, pH- and temperature-responsive hydrogels from crosslinked triblock copolymers prepared via consecutive atom transfer radical polymerizations. Biomaterials **27**, 2787–2797 (2006). https://doi.org/10.1016/j.biomaterials.2006.01.003
108. D. Han, Z. Lu, S.A. Chester, H. Lee, Micro 3D printing of a temperature-responsive hydrogel using projection micro-Stereolithography. Sci. Rep. **8**, 1963 (2018). https://doi.org/10.1038/s41598-018-20385-2

109. J. Lai, X. Ye, J. Liu, C. Wang, J. Li, X. Wang, M. Ma, M. Wang, 4D printing of highly printable and shape morphing hydrogels composed of alginate and methylcellulose. Mater. Des. **205**, 109699 (2021). https://doi.org/10.1016/j.matdes.2021.109699
110. M. Nizioł, J. Paleczny, A. Junka, A. Shavandi, A. Dawiec-Liśniewska, D. Podstawczyk, 3D printing of thermoresponsive hydrogel laden with an antimicrobial agent towards wound healing applications. Bioengineering **8**, 79 (2021). https://doi.org/10.3390/bioengineering8060079
111. X. Wang, Y. Yu, C. Yang, L. Shang, Y. Zhao, X. Shen, Dynamically responsive scaffolds from microfluidic 3D printing for skin flap regeneration. Adv. Sci. **9** (2022). https://doi.org/10.1002/advs.202201155
112. T. Abdullah, O. Okay, 4D printing of body temperature-responsive hydrogels based on poly(acrylic acid) with shape-memory and self-healing abilities. ACS Appl. Bio Mater. **6**, 703–711 (2023). https://doi.org/10.1021/acsabm.2c00939
113. P. Gupta, K. Vermani, S. Garg, Hydrogels: from controlled release to pH-responsive drug delivery. Drug Discov. Today **7**, 569–579 (2002). https://doi.org/10.1016/S1359-6446(02)02255-9
114. A.H. Choi, *Biomaterials and Bioceramics—Part 2: Nanocomposites in Osseointegration and Hard Tissue Regeneration* (2022), pp. 47–88. https://doi.org/10.1007/978-981-16-7435-8_2
115. C. Garcia, A. Gallardo, D. López, C. Elvira, A. Azzahti, E. Lopez-Martinez, A.L. Cortajarena, C.M. González-Henríquez, M.A. Sarabia-Vallejos, J. Rodríguez-Hernández, Smart pH-responsive antimicrobial hydrogel scaffolds prepared by additive manufacturing. ACS Appl. Bio Mater. **1**, 1337–1347 (2018). https://doi.org/10.1021/acsabm.8b00297
116. F. Wang, L. Li, X. Zhu, F. Chen, X. Han, Development of pH-responsive Polypills via semi-solid extrusion 3D printing. Bioengineering **10**, 402 (2023). https://doi.org/10.3390/bioengineering10040402
117. Y. Choi, C. Kim, H.S. Kim, C. Moon, K.Y. Lee, 3D printing of dynamic tissue scaffold by combining self-healing hydrogel and self-healing ferrogel. Colloids Surf. B Biointerfaces **208**, 112108 (2021). https://doi.org/10.1016/j.colsurfb.2021.112108
118. C.U. Mun, H.S. Kim, M. Kong, K.Y. Lee, Three-dimensional printing of hyaluronate-based self-healing ferrogel with enhanced stretchability. Colloids Surf. B Biointerfaces **221**, 113004 (2023). https://doi.org/10.1016/j.colsurfb.2022.113004
119. L. Vítková, N. Kazantseva, L. Musilová, P. Smolka, K. Valášková, K. Kocourková, M. Humeník, A. Minařík, P. Humpolíček, A. Mráček, I. Smolková, Magneto-responsive hyaluronan hydrogel for hyperthermia and bioprinting: magnetic, rheological properties and biocompatibility. APL Bioeng. **7** (2023). https://doi.org/10.1063/5.0147181
120. T.S. Tran, R. Balu, S. Mettu, N. Roy Choudhury, N.K. Dutta, 4D printing of hydrogels: innovation in material design and emerging smart systems for drug delivery. Pharmaceuticals **15**, 1282 (2022). https://doi.org/10.3390/ph15101282
121. P. Bender, J. Fock, M.F. Hansen, L.K. Bogart, P. Southern, F. Ludwig, F. Wiekhorst, W. Szczerba, L.J. Zeng, D. Heinke, N. Gehrke, M.T.F. Díaz, D. González-Alonso, J.I. Espeso, J.R. Fernández, C. Johansson, Influence of clustering on the magnetic properties and hyperthermia performance of iron oxide nanoparticles. Nanotechnology **29**, 425705 (2018). https://doi.org/10.1088/1361-6528/aad67d
122. H. Chu, W. Yang, L. Sun, S. Cai, R. Yang, W. Liang, H. Yu, L. Liu, 4D printing: a review on recent progresses. Micromachines (Basel) **11**, 796 (2020). https://doi.org/10.3390/mi11090796
123. M.C. Mulakkal, R.S. Trask, V.P. Ting, A.M. Seddon, Responsive cellulose-hydrogel composite ink for 4D printing. Mater. Des. **160**, 108–118 (2018). https://doi.org/10.1016/j.matdes.2018.09.009
124. S. Bashir, M. Hina, J. Iqbal, A.H. Rajpar, M.A. Mujtaba, N.A. Alghamdi, S. Wageh, K. Ramesh, S. Ramesh, Fundamental concepts of hydrogels: synthesis, properties, and their applications. Polymers (Basel) **12**, 2702 (2020). https://doi.org/10.3390/polym12112702
125. C. Lv, X.-C. Sun, H. Xia, Y.-H. Yu, G. Wang, X.-W. Cao, S.-X. Li, Y.-S. Wang, Q.-D. Chen, Y.-D. Yu, H.-B. Sun, Humidity-responsive actuation of programmable hydrogel microstructures

based on 3D printing. Sens Actuat. B Chem **259**, 736–744 (2018). https://doi.org/10.1016/j.snb.2017.12.053
126. L. Hiendlmeier, F. Zurita, J. Vogel, F. Del Duca, G. Al Boustani, H. Peng, I. Kopic, M. Nikić, T.F. Teshima, B. Wolfrum, 4D-printed soft and stretchable self-folding cuff electrodes for small-nerve interfacing. Adv. Mater. **35** (2023). https://doi.org/10.1002/adma.202210206
127. G.H. Yang, W. Kim, J. Kim, G. Kim, A skeleton muscle model using GelMA-based cell-aligned bioink processed with an electric-field assisted 3D/4D bioprinting. Theranostics **11**, 48–63 (2021). https://doi.org/10.7150/thno.50794
128. Y.-C. Li, Y.S. Zhang, A. Akpek, S.R. Shin, A. Khademhosseini, 4D bioprinting: the next-generation technology for biofabrication enabled by stimuli-responsive materials. Biofabrication **9**, 012001 (2016). https://doi.org/10.1088/1758-5090/9/1/012001
129. J.D. Finan, F. Guilak, The effects of osmotic stress on the structure and function of the cell nucleus. J. Cell. Biochem. **109**, 460–467 (2010). https://doi.org/10.1002/jcb.22437
130. Q. Ramadan, M. Zourob, 3D bioprinting at the frontier of regenerative medicine, pharmaceutical, and food industries. Front. Med. Technol. **2** (2021). https://doi.org/10.3389/fmedt.2020.607648
131. J. Malda, J. Visser, F.P. Melchels, T. Jüngst, W.E. Hennink, W.J.A. Dhert, J. Groll, D.W. Hutmacher, 25th anniversary article: engineering hydrogels for biofabrication. Adv. Mater. **25**, 5011–5028 (2013). https://doi.org/10.1002/adma.201302042
132. I.T. Ozbolat, Y. Yu, Bioprinting toward organ fabrication: challenges and future trends. I.E.E.E. Trans. Biomed. Eng. **60**, 691–699 (2013). https://doi.org/10.1109/TBME.2013.2243912
133. H.J. Lee, Y.B. Kim, S.H. Ahn, J. Lee, C.H. Jang, H. Yoon, W. Chun, G.H. Kim, A new approach for fabricating collagen/ECM-based bioinks using Preosteoblasts and human adipose stem cells. Adv. Healthc. Mater. **4**, 1359–1368 (2015). https://doi.org/10.1002/adhm.201500193
134. J. Groll, J.A. Burdick, D.-W. Cho, B. Derby, M. Gelinsky, S.C. Heilshorn, T. Jüngst, J. Malda, V.A. Mironov, K. Nakayama, A. Ovsianikov, W. Sun, S. Takeuchi, J.J. Yoo, T.B.F. Woodfield, A definition of bioinks and their distinction from biomaterial inks. Biofabrication **11**, 013001 (2018). https://doi.org/10.1088/1758-5090/aaec52
135. V. Mironov, Printing technology to produce living tissue. Expert. Opin. Biol. Ther. **3**, 701–704 (2003). https://doi.org/10.1517/14712598.3.5.701
136. Ž.P. Kačarević, P.M. Rider, S. Alkildani, S. Retnasingh, R. Smeets, O. Jung, Z. Ivanišević, M. Barbeck, An introduction to 3D bioprinting: possibilities. Chall.S Futur. Asp. Mater. **11**, 2199 (2018). https://doi.org/10.3390/ma11112199
137. J. Li, A.D. Celiz, J. Yang, Q. Yang, I. Wamala, W. Whyte, B.R. Seo, N.V. Vasilyev, J.J. Vlassak, Z. Suo, D.J. Mooney, Tough adhesives for diverse wet surfaces. Science **357**(2017), 378–381 (1979). https://doi.org/10.1126/science.aah6362
138. S.V. Murphy, A. Atala, 3D bioprinting of tissues and organs. Nat. Biotechnol. **32**, 773–785 (2014). https://doi.org/10.1038/nbt.2958

Chapter 6
Theranostic Applications of Hydrogels and Ferrogels: From Laboratory to Clinics

6.1 Hydrogels and Ferrogels in the Treatment of Diabetes

High blood glucose is an indication of diabetes, a chronic metabolic condition. Traditional diabetic treatment involves subcutaneous injections and skin punctures, which are very inconvenient for diabetic patients. Due to their unique properties, including stretchability, excellent biocompatibility, and biodegradability, HG-FGs provide versatile and effective materials for diabetes diagnosis and therapy [1, 2]. Patients with diabetes often see fluctuations in their blood sugar levels as a result of either insulin resistance or insufficient insulin production. Diagnosing and treating diabetes requires routine blood glucose testing. Wearable HG-FGs materials that combine diagnostic and therapeutic capabilities might significantly reduce the pain of routine testing for diabetic patients. Three main tenets now underpin continuous blood glucose monitoring: Initially, phenylboronic acid (PBA) fragments are used to bond with glucose. Second, glucose oxidase (GOx) catalyzes the hydrolysis of glucose into hydrogen peroxide and d-glucose lactone, which in turn hydrolyzes into gluconic acid. Third, concanavalin A (ConA) binds to glucose because to its high affinity for hydrocarbon groups [3].

Because of its exceptional biocompatibility and degradability, PBA has been developed for glucose sensing by adding it to HG-FGs. Yetisen and his team [4] developed a hydrogel-based optical fiber material consisting of a polyacrylamide-co-polyethylene glycol diacrylate core and a calcium alginate coating. For sensing purposes, this material makes use of the PBA-glucose binding principle. The hydrogel's density changes as a result of an increase in osmotic pressure brought on by glucose's interaction with PBA and entry into the core. By monitoring changes in refractive index (RI), one may quantify the displacement of light propagation inside the optical fiber caused by this. This enables quantitative testing for glucose levels. Liang et al. developed a glucose/pH dual-responsive hydrogel dressing for diagnosis and treatment in 2022 using Schiff base and boronate ester as the dual dynamic bond system [5]. The hydrogel's exceptional self-healing capabilities were

S. Awasthi, *Smart Hydrogels and Ferrogels*, Materials Horizons: From Nature to Nanomaterials, https://doi.org/10.1007/978-981-95-9077-3_6

made possible by the existence of dual dynamic bonds. While the catechol structure enhanced the hydrogel's better blood clotting, tissue adhesion, in vivo hemostatic capabilities, antioxidant qualities, and biocompatibility, L-arginine supplied antibacterial characteristics. To lessen inflammation and encourage angiogenesis, when glucose abnormalities occur, the hydrogel releases graphene oxide (GO) and metformin (Met).

As a complex pathophysiological process, diabetic foot ulcers (DFU) currently have no known cure. The unique wound environment of diabetic wounds is created by poor angiogenesis, hyperglycemia, neuropathy, advanced glycation end products, and reactive oxygen species (ROS). The goal of the work done by Emad et al. [6] was to create a topical treatment for DFU using naringenin-ferulic acid beeswax-based dual-responsive hydrogel (NAR-FA NLC HG). Hydrogel was compatible with HaCaT cells, according to cell line research, and after 48 h of scratch wound assay, wound contraction was 92.4%. In STZ-induced diabetic rats, the rate of wound closure with NAR FA NLC HG was 93.2% at day 15 (Fig. 6.1i A, B). According to Fig. 6.1i C, the course of wound healing over time, the groups treated with NAR gel, NAR-FA NLC gel, and NAR-FA NLC HG had better wound repair outcomes than the diabetes control group. It was discovered that the wound dressings' hydrogel (dual-responsive) and beeswax-based NAR-FA NLC enhanced wound re-epithelialization, which might account for the wound's quicker healing compared with the other treatments. With its built-in components of antibacterial activity, such as beeswax, NAR, and FA, the smart hydrogel developed in this work might thus be a viable topical treatment for DFU.

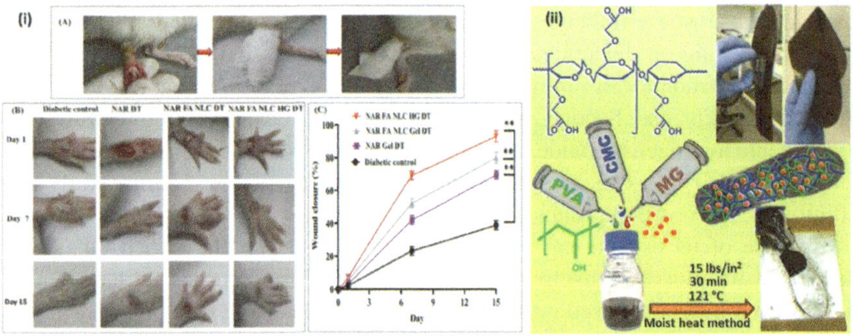

Fig. 6.1 **i** **A** Creation of wounds and application of dressings in a diabetic rat model. **B** Monitoring of wound healing progression following treatment with phosphate-buffered saline (PBS, diabetic control), naringenin-loaded nanostructured lipid carrier gel (NAR DT), naringenin–ferulic acid-loaded nanostructured lipid carrier gel (NAR-FA NLC DT), and a dual-responsive hydrogel containing Naringenin–Ferulic acid nanostructured lipid carriers (NAR-FA NLC HG DT) on days 1, 7, and 15 post-injury. **C** Comparison of wound closure percentages across different treatment groups. Reproduced with permission from Ref. [6]. Copyright 2025 Elsevier. **ii** Diagram to represent polyvinyl alcohol/carboxymethyl cellulose hydrogels as effective diabetic shoe insoles. Reproduced with permission from Ref. [9]. Copyright 2020 Elsevier

Insulin resistance or β-cell loss impairs the body's capacity to create enough insulin, which disrupts glucose homeostasis and causes a sharp increase in blood glucose levels. In order to prevent repercussions like organ degeneration and to maintain normal blood glucose levels, diabetes individuals need to take medication, ideally by simulating normal insulin production. The usual and final course of therapy is regular subcutaneous insulin injections to maintain the natural glucose balance. Insulin can be delivered steadily and continuously to address these problems. In this regard, cyclodextrin extended polyurethane (PU-co-CD)/CMC hydrogel (injectable) was developed to administer insulin continuously over a long period of time to control blood glucose [7]. The insulin-encumbered PU-co-CD copolymer's special composition allows the carboxymethyl cellulose (CMC) to create an injectable hydrogel. Spectroscopic, chromatographic, and rheological analyses have confirmed the manufacturing and processing of this chain-extended copolymer. Insulin release in diabetic Swiss albino mice is controlled for up to three days by insulin encapsulated in PU-co-CD administered subcutaneously in a gel medium, as opposed to less than twelve hours with traditional insulin injection. Because it is biocompatible and biodegradable, PU-co-CD-gel is a viable drug delivery medium. Histopathology showed that the formulation provides prolonged insulin delivery in diabetes conditions without causing damage to major organs.

Furthermore, Xing and the group [8] described how a bone-adhesive hydrogel might be made and injected locally into the fracture site, and that, in reaction to ROS activity (ROS), releases rutin, a natural inhibitor of nod-like receptor, protein domain-associated protein 3 (NLRP3). Poly(vinyl alcohol), Rutin-functionalized gelatin, and oxidized starch were readily cross-linked to form the hydrogel (known as RPO) based on the dynamic Schiff base and boronate ester linkage. In the RPO hydrogel, rutin is conjugated via a ROS linker and then released when active ROS cleave the linker. The osteogenic development of bone marrow mesenchymal stem cells is promoted by RPO hydrogel, which significantly reduces mitochondrial dysfunction, oxidative stress, and the overactivation of the NLRP3 inflammasome in bone marrow-derived macrophages, according to in vitro experiments. By altering the inflammatory milieu, RPO hydrogel enhanced bone regeneration in a fracture model in diabetic rats. The findings showed that employing RPO hydrogel, an injected bioresponsive adhesive to treat diabetes fractures, can lead to local, on-demand NLRP3 suppression.

Nowadays, shoe inserts made of plastic, rubber, or silicone are utilized as a preventative measure against diabetic foot. These inserts are extremely painful because of their non-biodegradable, non-absorbent nature, and magnetic protrusions. In addition, their disposal after use contributes to environmental pollution. Therefore, the purpose of the study carried out by Patwa and the group was to assess magnetic hydrogels based on polyvinyl alcohol/carboxymethyl cellulose (PVA/CMC) that were made by physical crosslinking with varying concentrations of strontium ferrite nanoparticles (MG) as a substitute for diabetic shoe inserts (Fig. 6.1ii). It was noted that the addition of MG improved the overall characteristics. CMC and PVA's strong hydrogen bonding was discovered using infrared spectroscopy. The results demonstrated that adding MG increased the PVA/CMC hydrogels' tensile strength and flexibility by around 20% and 40%, respectively. Furthermore, in order to prevent the growth of microorganisms on skin, the ferrogels were able to absorb around 300% of their initial weight in moisture. For diabetic shoe insoles, PVA/CMC/MG gels can therefore be regarded as a biodegradable substitute.

6.2 HG-FGs in Bone Tissue Engineering

Bone tissue damage recovery includes mineralization, bone remodeling, early hemostasis, and an immediate inflammatory response [10–12]. High blood sugar constantly stimulates the release of insulin by the pancreatic beta cells in addition to raising the patients' blood glucose levels. Long-term stimulation, however, may result in beta cell malfunction, which exacerbates diabetes. Furthermore, changes in blood sugar can cause metabolic and mitochondrial malfunction, which can be detrimental to tissue regeneration [13]. ROS can be produced as a result of glucose level variations, which may experienced by diabetic patients. At the site of bone abnormalities, these ROS may subsequently promote a pro-inflammatory environment at the site of bone defects. As a result, downstream osteoprogenitor cells become less functional, which impacts the bone-healing process. Thus, by creating a hydrogel (double-network) composed of gelatin colloids and polyvinyl alcohol that has been crosslinked with phenylboronic acid, Li et al. [14] were able to effectively establish a hydrogel method for bone healing in diabetics. There are two uses for this system: "diagnosis" and "treatment." By examining pathogenic indicators such as ROS, growth factors, and matrix metalloproteinases, it decides when to provide drugs in the diabetic environment. It efficiently promotes bone tissue regeneration by achieving the slow release of many drugs to more effectively support the immunological and bone-healing cascade.

Awasthi and the co-workers [15] designed a study related to polyacrylamide (PAM) hydrogels with various concentrations of hexagonal boron nitride (hBN) and silver (Ag) for cartilage repair. PAM-Ag1, PAM-BN3, and PAM-Ag-BN's effects on the cytotoxicity of chondrocyte C28/I2 cells are shown in Fig. 6.2A. For 3, 5, and 7 days, PAM-BN3 showed cell viability of 97.4%, 97.1%, and 96.4%, respectively, whereas PAM-Ag1 demonstrated 95.8%, 95.5%, and 94.5%. Hexagonal boron nitride functionalization using the OH group was the cause of the increased cell viability in PAM-BN3. For 3, 5, and 7 days, PAM-Ag-BN showed 98.5%, 98%, and 97.3% cell viability, respectively. Ag and functionalized BN worked in concert to increase the PAM-Ag-BN trimer composite's cell viability percentage. Furthermore, during the course of incubation, MTT readings reveal non-significant changes between PAM-Ag1, PAM-BN3, and PAM-Ag-BN (Fig. 6.2A d). Over the course of 3, 5, and 7 days, the OD values at 570 nm increased, demonstrating the proliferation of chondrocyte C28/I2 cells in all hydrogel samples. Appropriate bone replacements are desperately needed for patients with significant bone abnormalities. In addition to being biocompatible, these substitutes must resemble the three-dimensional structure of bone and have comparable mechanical, chemical, and biological characteristics, ideally encouraging ossification. Since collagen type I and carbonate hydroxyapatite make up the majority of real bone, a biomaterial that can be 3D printed using nanohydroxyapatite (nHAp) and methacrylated gelatin (GelMA) would be useful for simulating the composition and structure of genuine bone. Until now, these nanocomposite hydrogels (NCH) have poor rheological characteristics that have prevented them from being used for high-structural-fidelity extrusion-based 3D printing. Naoloue et al. [16] provided a new GelMA/nHAp NCH composition that addresses the problem of calcium cations produced by nHAp that prevent GelMA gelation and incorporates the rheological modifier carbomer to enhance rheological capabilities. The printed scaffolds demonstrated good cell survival and osteogenic differentiation, highlighting their potential for bone tissue engineering.

Sousa et al. [17] were the first to create a stiff three-dimensional structure using hydroxyapatite nanoparticles (nHA) and polycaprolactone (PCL) combined with a bioink that contains collagen (Col), alginate, nHA, and human dental pulp stem/stromal cells (hDPSCs). Through cytocompatibility testing, osteogenic differentiation evaluation, and biocompatibility assessment in a rat model, the biofabricated constructions were thoroughly described. After 168 h, the hybrid scaffolds showed noticeably greater vitality of cells compared to the control group (in vitro). In addition, compared to other groups, the hybrid scaffolds had higher osteogenic differentiation. Good biocompatibility was demonstrated by the in vivo study, which was marked by a successful tissue integration and a low inflammatory reaction. By combining the biological activities of hDPSCs bioink and alginate-nHA-Col with the mechanical strength of PCL and nHA, the results demonstrate the scaffold's ability to promote bone tissue regeneration. The present work provided a promising foundation for the creation of biomaterials that will enhance clinical results in bone repair and regeneration for the treatment of critical-size bone lesions, targeted drug delivery, and 3D models for bone tissue engineering.

Normal physiological processes depend heavily on ROS. However, oxidative stress due to hypoxia causes an excessive generation of ROS in the bone defect microenvironment, which may disturb tissue healing. Therefore, encouraging bone regeneration requires controlling ROS. The current research described the design and preparation of a mesoporous bioactive glass-based nanoenzyme that has the ability to control ROS (MBGNs@PDA@MnO_2) [18]. The MBGNs@PDA@MnO_2 demonstrated improved mineralization characteristics and antioxidation behavior. By utilizing oxidized chitosan and dopamine-grafted gelatin (OCS and Gel-DA, respectively) through an oxidative coupling and Schiff-base process, a double-network hydrogel was created. MBGNs@PDA@MnO_2 was then added to the hydrogel. Cell tests showed high cytocompatibility, the capacity to promote osteogenesis, and the ability to efficiently lower intracellular ROS levels by the fabricated hydrogels. Therefore, the ROS-regulated composite hydrogels held promise as a novel approach to treating bone deficiencies (Fig. 6.2B).

Ferrogels' potential to improve bone regeneration and healing is being investigated in bone tissue engineering. Cells can be directly encapsulated in the ferrogel

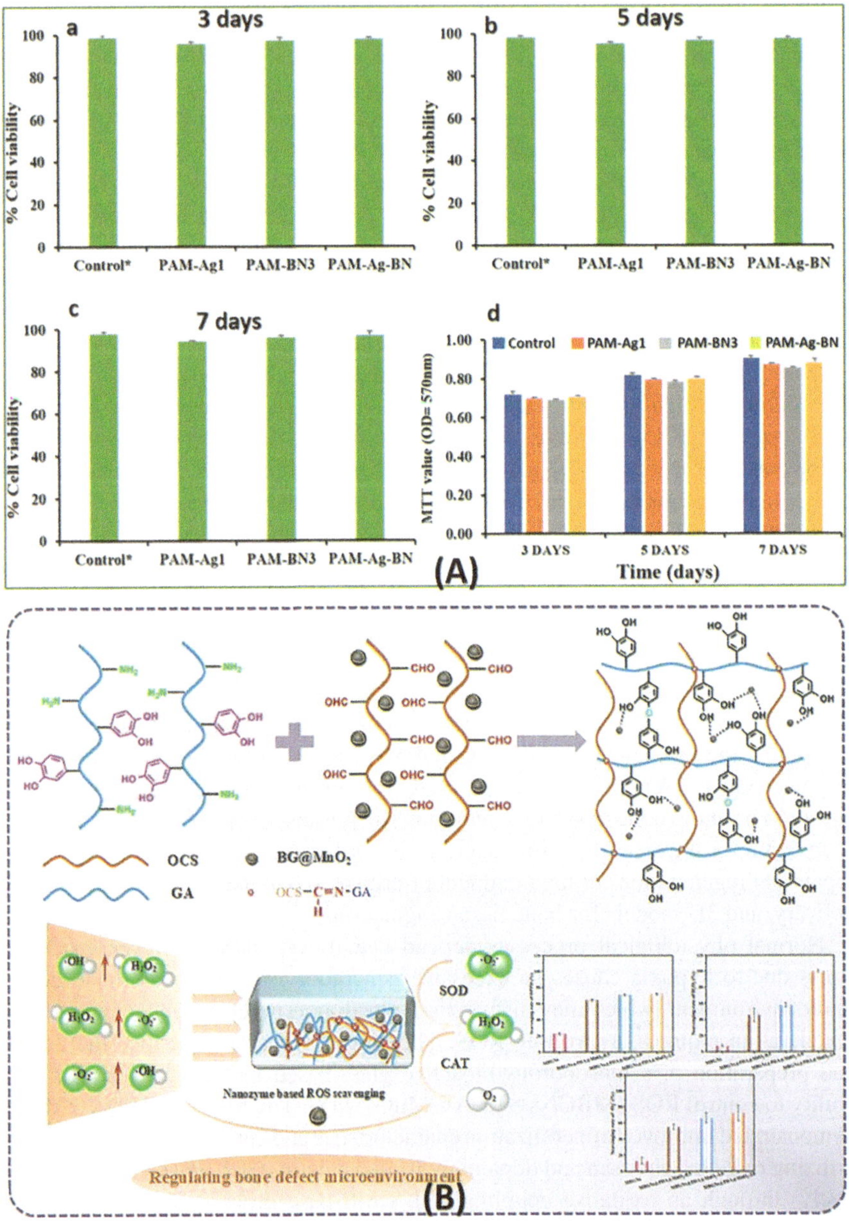

Fig. 6.2 **A** Effects of PAM-Ag1, PAM-BN3, and PAM-Ag-BN on the viability of chondrocyte C28/I2 cells for 3 (**a**), 5 (**b**), and 7 (**c**) days; **d** MTT values over 3, 5, and 7 days with PAM-Ag1, PAM-BN3, and PAM-Ag-BN in chondrocyte C28/I2 cells at OD 570 nm. Reproduced with permission from Ref. [15]. Copyright 2025 Royal Society of Chemistry. **B** Mechanisms of crosslinking and ROS removal in composite hydrogels for bone tissue engineering. Reproduced with permission from Ref. [18]. Copyright 2025 Elsevier

substance or utilized as a distinct phase. Ferrogels may also be magnetically activated to control cells and enhance bone growth. Because of its excellent biocompatibility, biodegradability, and designability, poly(γ-glutamic acid) (γ-PGA) hydrogels have been extensively researched and used in the biomedical area. However, the low mechanical strength, lack of integrity with surrounding tissues, and acidic environment caused by γ-PGA degradation restrict the biomedical uses of γ-PGA hydrogels. Fe-doped mesoporous bioglass nanospheres (Fe-MBG NPs) were initially produced in order to overcome these mentioned restrictions. Fe-MBG NPs-composited γ-PGA (Fe-MBG/γ-PGA) hydrogels were then created by Gao and the team [19]. According to the results, Fe-MBG NPs with a typical diameter of 426 ± 8 nm were effectively produced. The 30%Fe-MBG/γ-PGA ferrogels that were created demonstrated good bioactivity, conductivity, and cytocompatibility in addition to significant injectability, adhesion, stretchability, elasticity, and fatigue resistance. Consequently, the 30%Fe-MBG/γ-PGA ferrogels showed good potential for use in bone tissue regeneration.

6.3 Cancer Therapy

The use of hydrogels in cancer treatment has attracted a lot of attention in recent years. In particular, stimulus-responsive hydrogels and ferrogels have become well-known platforms for precisely regulating anti-cancer drug delivery. By responding to certain stimuli, such as temperature, pH, enzymes, or light, these HG-FGs deliver therapeutic medicines to tumor tissues in a targeted and sustained manner. The three-dimensional structure of tumor tissues can be replicated using HG-FGs, providing a more realistic environment for studying tumour cell proliferation, invasion, and metastasis [20]. By introducing tumor cells into HG-FGs, researchers may investigate the biological activities of tumors and how they respond to drugs. However, because of their highly changeable physical and chemical properties, HG-FGs may be employed as anti-cancer drug delivery systems that load and release drugs. Pharmaceuticals may work with HG-FGs to offer targeted distribution and delayed release by adsorption, encapsulation, or covalent bonding, increasing therapeutic efficacy and reducing side effects. Compared to traditional hydrogels, stimulus-responsive HG-FGs provide better control over the local and sustained release of medications. One example is the use of thermoresponsive hydrogels, which show a sol–gel transition exactly where the tumor is located. These HG-FGs may be readily injected while still soluble and subsequently solidified in situ to form implants. Unlike pre-formed HG-FGs, these implants act as reservoir systems by conforming to the surrounding tissue, extending the period of drug release [21]. Thus, a potential trend in the future of HG-FGs-based cancer therapy is the creation of multifunctional HG-FGs that can react to the tumor microenvironment for drug release and detection.

It is difficult for proteins, which are naturally biocompatible and biodegradable, to produce stable hydrogels without the use of chemical crosslinkers. Eom et al. [22] created a core protein building block called SpyTag-fused proliferating cell nuclear antigen protein (ST-PCNA) in the form of a ring and a dumbbell-shaped tandem dimeric SpyCatcher (SC-SC) as a bridge component. By mixing the ST-PCNA and SC-SC solutions, self-crosslinked PCNA/SC-SC protein (2SP) hydrogels may be successfully produced without the need for chemical crosslinkers. Different cargo molecules, such as anti-cancer medications, functional proteins, and photo-sensitizers, are effectively integrated during their creation by mixing, resulting in cargo@2SP hydrogels. Through localized injection beneath the 4 T1 tumor in mice, 2SP hydrogels containing anti-cancer drugs or photosensitizers have been successfully created. Tumor development is significantly suppressed by the targeted, progressive release of medications in the physiological milieu and by the extremely localized photosensitizers that are preserved in the 2SP hydrogels, which elevate the local temperature over 45 °C when exposed to laser light. Furthermore, topical application of 2SP hydrogels containing growth factor proteins to the incision wound region promotes fast extracellular matrix rebuilding and efficient skin regeneration. The self-crosslinkable and injectable 2SP hydrogels created here have potential as innovative biocompatible scaffolds for local treatment (Fig. 6.3A).

5-Fluorouracil (5-FU), a drug often used in cancer treatment, has promising results in colon cancer treatment. Although 5-FU is commonly administered via injection into a vein, oral rate-controlled formulations have been shown to reduce the adverse effects [23]. To produce controlled drug delivery systems, several studies have encapsulated 5-FU in polymeric materials. The increased drug efficiency of controlled drug release systems has received a lot of interest in the present day [24, 25]. A potential treatment approach for advanced colorectal cancer (CRC) is the combination of photothermal therapy (PTT) and chemotherapy delivered via hydrogel [25]. Significant difficulties arise, though, in preserving the stability of the photo-absorbing agents (PTA) in the hydrogel and making sure the hydrogel stays tethered to the tumor tissue. In particular, N-(2-aminoethyl)-4-(4-(hydroxymethyl)-2-methoxy-5-nitro-sophenoxy) butanamide-linked sodium alginate-MoS_2-5-fluorouracil (AlgNB/MoS_2/5-FU) is an injectable 2D molybdenum disulfide (2D-MoS_2)-integrated adhesive hydrogel. This hydrogel treats colorectal cancer (CRC) by serving as a drug-delivery and photothermal implant that is activated by near-infrared light (NIR). The MoS_2 nanosheets have a very effective NIR-triggered photothermal action and retain good dispersibility in the hydrogel. Crucially, AlgNB's aldehyde group also gave the hydrogel tissue adhesion; this sticky hydrogel is employed to penetrate and adhere to tumor tissue. In both in vitro and in vivo testing, the injectable adhesive AlgNB/MoS_2/5-FU hydrogel demonstrates exceptional performance in promoting colorectal tumor regression and reducing the proliferation of SW480 cells by inducing PTT and administering the 5-FU medication. A strong immune response may be induced, and DNA repair inhibited by the potential synergistic mechanism between 5-FU and PTT treatment. As a result, this study offers insightful information for the cooperative management of localized colorectal cancer.

Because injectable hydrogels provide prolonged, controlled drug administration at specific tumor locations and minimize the adverse consequences of the drug's systemic exposure, intratumor local drug delivery devices have drawn a lot of attention. This work concentrated on photodynamic/photothermal therapy (PDT and PTT,

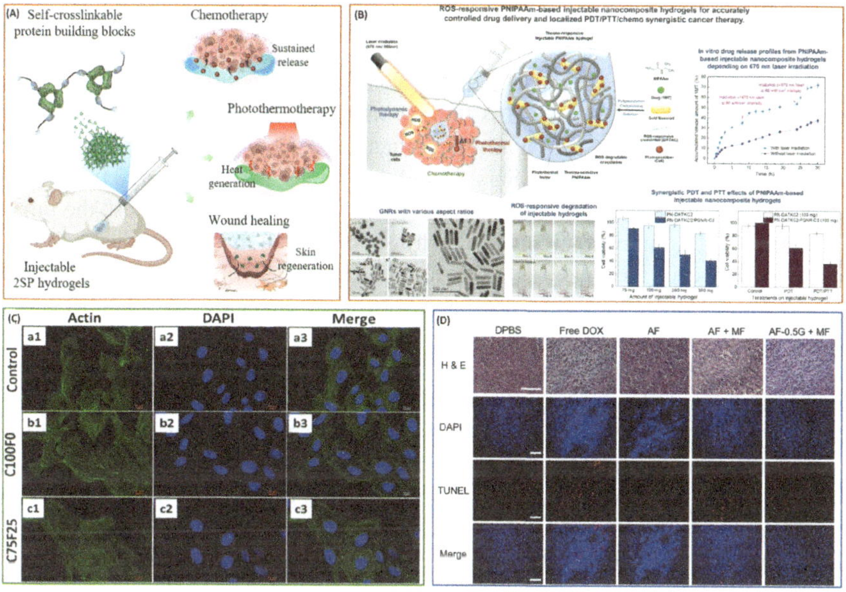

Fig. 6.3 **A** Injectable 2SP hydrogels, which are injectable and self-crosslinkable, have the potential for cancer therapy and wound healing. Reproduced with permission from Ref. [25]. Copyright 2025 Elsevier. **B** ROS-responsive PNIPAAm injectable hydrogels for controlled drug release and photodynamic treatment (PDT)/photothermal therapy (PTT) synergistic cancer therapy. Reproduced with permission from Ref. [26]. Copyright 2025 Elsevier. **C** Actin filaments were labeled with Alexa Fluor-488 primary antibody, and nuclei with DAPI in the cytoskeleton pictures of **a1**–**a3** control HOS cells, **b1**–**b3** cells treated with C100F0, and **c1**–**c3** cells treated with C75F25 at a 50 μg mL^{-1} concentration for 24 h. The images were scaled to 10 μm. Reproduced with permission from Ref. [39]. Copyright 2022 Elsevier. **D** Tumour tissues obtained from mice treated with DOX-containing alginate ferrogel were stained with H&E and TUNEL (scale bar, 100 μm). Reproduced with permission from Ref. [40]. Copyright 2019 Elsevier

respectively) coactive combination treatment and cytotoxic ROS-cleavable injectable hydrogels that regulate drug delivery more precisely to enhance cancer therapeutic efficiency. Injectable hydrogels based on thermoresponsive poly(N-isopropyl acrylamide) (PNIPAAm) were created with a crosslinker (thioketal) that is susceptible to bond breaking by ROS [26]. With a lower critical solution temperature (LCST) between 33.6 and 36.3 °C, the injectable nanocomposite hydrogel based on PNIPAAm could be injected at ambient temperature using a syringe. Nevertheless, a phase change at 37 °C body temperature produced a gel state. Photosensitizer (PS)-conjugated GNR feed quantity and 670 nm laser intensity were adjusted to regulate ROS production from PNIPAAm-based hydrogel (injectable). These hydrogels demonstrated selective degradation and drug release behavior, crosslinked with ROS-cleavable thioketal crosslinker. Furthermore, under laser irradiation, a PNIPAAm hydrogel with gold nanorods and chlorin e6 (Ce6) (as a photothermal factor and as a PS, respectively) demonstrated a PDT/PTT synergistic effect. According to these results, this ROS-sensitive nanocomposite hydrogel will provide a precise drug delivery platform and an effective combinational cancer therapy system of tailored PDT and PTT (Fig. 6.3B).

One of the most prevalent and deadly cancers in the world, breast cancer has a high death rate [27]. Although it is the recommended course of therapy for local breast tumors, surgical resection has a significant risk of surgical complications and inadequate tumor removal [28]. Surgery-induced metastasis can emerge from them, since they can promote many pathophysiological mechanisms that cause the tumor to return [29, 30]. Intravenous anticancer drug administration is part of the traditional treatment; however, it is linked to systemic toxicity, drug distribution to major organs, Poor treatment outcomes, drug resistance, restricted medication availability at the tumor site, and metastases of malignancy [31, 32]. Therefore, to avoid postoperative cancer recurrence, techniques for optimizing medication availability at the tumor site are necessary. Accordingly, localized drug delivery methods have been found to be more successful in reducing systemic side effects by decreasing organ diffusion and sustaining drug concentration at the target region for an extended length of time [33, 34]. Many researchers have recently reported using injectable hydrogels for cancer treatment and localized medication delivery [35, 36]. For a variety of anticancer drug delivery applications, injectable hydrogels displaying a sol–gel transition at body temperature have been documented. Paclitaxel nanocrystals that are lipid-coated, Soluplus stabilized, and modified with fucoidan-oleylamine conjugate are part of the study carried out by Kumar et al. [37]. When compared to pure paclitaxel (PTX), the homogenous, stable, and around 100 nm-long nanocrystals (Lipid-NCs) demonstrated better drug release. For the intratumorally injection, the nanocrystals were then placed into an in-situ gel-forming hydrogel. The resulting hydrogel showed a sol-form at lower temperatures (between 2 and 8 °C); above body temperature, it changed to a gel-form. The injectable hydrogel demonstrated excellent syringeability, sufficient viscosity, acceptable pH, and a quick sol–gel transition. The hydrogel demonstrated a high payload potential, consistent dispersion, and controlled extended release of medications. Lipid-NCs hydrogel outperformed both pure PTX-loaded hydrogel and intravenously delivered PTX in in vivo tests for tumor suppression without systemic harm. In conclusion, medication for breast cancer may be administered locally using a hydrogel containing nanocrystals.

Ferrogels, in addition to hydrogels, are promising options for cancer treatment because of their combination of magnetic responsiveness and biocompatibility. Under magnetic direction, ferrogels allow therapeutic chemicals to be released at specified places, allowing for controlled and targeted medication delivery. Ferrogels may also be utilized in magnetic hyperthermia, a process that produces localized heat to destroy cancer cells without causing harm to nearby healthy tissue. They are advanced cancer therapy techniques because of their multifunctionality. A magnetic nanocomposite hydrogel bead that is pH-sensitive and biocompatible was created as a possible oral medication delivery system [24]. In situ synthesis of Fe_3O_4 nanoparticles (NPs) with hydroxyapatite (HAp) as a substrate was followed by the incorporation of the HAp-Fe_3O_4 composite into the sodium alginate (SA) hydrogel (pH-sensitive). The anticancer medication 5-fluorouracil (5-FU) was ultimately loaded into the SA/HAp-Fe_3O_4 with success. In acidic conditions (pH 1.2), the SA/HAp-Fe_3O_4 gel bead exhibits little swelling, but in alkaline conditions (pH 7.4), it exhibits considerable swelling. Furthermore, drug release trials showed that this drug delivery system could hold onto the medication in the stomach's acidic environment while releasing it into the intestine's higher pH. Kinetic study of the release profiles showed that Fickian diffusion, a mechanism regulated by diffusion without contact between the drug and the gel matrix, was mostly accountable for the delivery rate. Cell vitality exceeded 85% in 4–48 μg/mL ($P < 0.05$), indicating high cytocompatibility for the SA/HAp/Fe_3O_4 magnetic nanocomposite hydrogel bead, despite 5-FU-loaded SA/HAp-Fe_3O_4 exhibiting obvious toxicity against HT-29 human colon cancer cells. The generated magnetic nanocomposite hydrogel beads' cytotoxicity, controlled release, and pH sensitivity findings suggest that it might be used as a targeted oral drug delivery technique for the treatment of colon cancer.

Combinatorial therapy, which consists of magnetic hyperthermia and chemotherapy, may offer a viable cancer treatment option by improving the safety and efficacy of the medication. Verma and the research team [38] created a magnetic-hydroxyapatite nanocomposite with the intention of using it for chemo-hyperthermia treatment. The nanocomposite was created using a soft-chemical method in which phosphate-coated Fe_3O_4 nanoparticles served as the core, and chemical precipitation was utilized to bind hydroxyapatite to their surface. Gelatin moieties were added to the nanocomposite in-situ to provide active sites for drug binding and water colloidal stability. With a drug concentration of around 8.2%, the nanocomposite demonstrated a high payload of the anticancer medication doxorubicin hydrochloride (DOX). Under acidic pH circumstances, a significant release of the drug was seen, indicating a pH-responsive sustained release that is advantageous for cancer treatment. In human cancer cells (MCF-7 and A549), the cytotoxicity test indicated that the nanocomposite was non-toxic; nevertheless, when compared to free DOX, the nanocomposite loaded with DOX exhibited increased toxicity and cellular absorption in cancer cells. Under an AC magnetic field (AMF), the nanocomposite demonstrated superparamagnetic behavior with heating efficacy dependent on particle concentration and magnetic field strength. Under an AMF of 0.422 kOe, the nanocomposite's specific absorption rate values fell within the range of 100–300 W/g, which is necessary for treating magnetic hyperthermia. Overall, the study showed that the created nanocomposite may be used for hyperthermia treatment as both a heating source and a medication carrier.

The creation of ferrogels made of chitosan for magnetically driven drug delivery and the treatment of hyperthermia is demonstrated simply by Sumitha et al. [39]. They exhibit good biodegradation, tumour-specific remotely switchable drug release, strong magnetic hyperthermia potential (42 °C in 4 min), and high drug loading efficiency (~ 40%) for doxorubicin, or DOX. Under normal settings (pH = 7.4), the ferrogels degrade almost completely in 3 months, but in the tumour-specific microenvironment (pH = 5.6), the pace of breakdown is accelerated and takes around 8 weeks. Additionally, after 60 min of exposure to a 50 mT magnetic field, a 30% increase in medication release was noted. Ferrogels have demonstrated improved bioactivity and capable cell-material interactions, which have been confirmed by confocal imaging, cytoskeleton staining, and the MTT experiment. The action of the released medication has significantly decreased the survivability of DOX-loaded samples, confirming the effectiveness of DOX-loaded ferrogels. The human osteosarcoma (HOS) cell morphology was observed after 4′,6-diamidino-2-phenylindole (DAPI) and Alexa Flour 488 staining, which were designated for cell nuclei and F-actins, respectively, in order to have a better understanding of the cell-material interaction. Figure 6.3C provides the typical pictures. Therefore, the described system has several functions that allow for advanced cancer treatment.

In order to improve treatment effectiveness in a tumour-bearing mouse model, Kim et al. [40] postulated that a macroporous alginate ferrogel (AF) may be used to release anticancer medications on-demand while being stimulated by magnetic fields. Following magnetic stimulation, the in vitro cancer cells' viability, grown on DOX-delivering alginate ferrogel, was much lower than that of cells grown on alginate ferrogel. After magnetic stimulation, DOX-releasing AF's (macroporous) therapeutic effectiveness also rose in tumour-bearing animals. The therapeutic effectiveness of AF was further corroborated by histological examination of tissue slices taken from the mice (Fig. 6.3D). Mice treated with 0.5G ferrogel (with magnetic field, MF) showed significant tissue cavitation and apoptotic cell death in comparison to mice treated with AF + MF. The controlled release of DOX from the gel by magnetic stimulation might be the cause of this discovery. Therefore, this method of creating a ferrogel that responds to magnetic stimulation might be helpful in the creation of smart drug delivery systems.

6.4 Wound Healing by HG-FGs

Wound healing is a natural self-repair process that involves coagulation, inflammation, scar formation, and tissue remodeling [41, 42]. HG-FGs have become desirable alternatives for wound dressings because of their better wettability, biocompatibility, and 3D porous structure, mimicking the extracellular matrix. Because the process of wound healing is intricate, using HG-FGs for a single treatment may not be sufficient to address diverse clinical needs. This highlights the need for multifunctional HG-FG materials, which can release therapeutic agents in response to specific physiological conditions. The goal of this strategy is to provide more effective and focused therapies.

Chronic inflammation, poor angiogenesis, and cellular senescence brought on by hyperglycemia make diabetic wound healing a major therapeutic problem. In order to solve these problems, Yang et al. [43] created a multifunctional hydrogel (GelMA/PNS/Alg@IGF-1) that combines sodium alginate microspheres containing insulin-like growth factor-1 (IGF-1), gelatine methacryloyl (GelMA), and Panax notoginseng saponins (PNS). This hydrogel was designed to target ageing and encourage vascular regeneration by achieving a gradient and continuous release of bioactive chemicals. Through the inhibition of NF-κB pathway activation, the hydrogel was shown in vitro to considerably reduce oxidative stress, suppress senescence markers and senescence-associated secretory phenotypes, and restore endothelial cell function under high-glucose settings. Transcriptomic analysis showed that pathways related to angiogenesis, apoptosis, and inflammation were modulated. In a rat model, this hydrogel improved collagen deposition, neovascularization, and granulation tissue growth while speeding up diabetic wound regeneration in vivo. Additionally, the hydrogel improved tissue remodeling while reducing oxidative stress and cellular senescence. PNS and IGF-1 worked in concert inside the hydrogel to create a pro-regenerative milieu that addressed vascular dysfunction and pathological aging. These results demonstrated that GelMA/PNS/Alg@IGF-1 was a viable treatment platform for diabetic wound care because it provided proangiogenic and anti-senescence effectiveness to address the challenges associated with chronic wound healing (Fig. 6.4A).

It has been demonstrated that hydrogen sulfide (H_2S) speeds up wound healing by encouraging angiogenesis, collagen deposition, proliferation, and inflammation reduction. However, uncontrolled and excessive release is a typical problem with widely used H_2S donors. To solve this, α-phase manganese sulfide nanoparticles (MnS NPs) were encapsulated in an injectable, pH-responsive polysaccharide hydrogel to provide regulated, on-demand H_2S administration for better wound healing (Fig. 6.4B) [44]. The hydrogel, known as MnS@AC, was created via dynamic Schiff base crosslinking of carboxymethyl chitosan (CCS) with aldehyde hyaluronic acid (AHA). During the first phases of wound curing, the regulated release of MnS NPs was made easier by the hydrogel networks. The MnS@AC group had a greater wound closure rate on day 13 (94.2%) in the full-thickness skin wound models than the AC group, MnS group, and control group (81.1%, 81.7%, and 69.9%, respectively). Additional studies showed that MnS@AC hydrogels promoted collagen deposition, increased angiogenesis, proliferation, and decreased proinflammatory cytokines to aid wound healing. Manganese ions produced from MnS NPs prevented undesired inflammation by not triggering an inflammatory response mediated by the cGAS-STING pathway. The pH-responsive hydrogels have shown improved low-toxic wound dressings that provide a new method of wound healing.

Dopamine-grafted polyacrylamide/sodium alginate/gelatin grafted with glycidyl methacrylate doped with polysaccharide from Angelica sinensis (PDGA) was made in a recent work by Liu and the team [45]. The ability of the PDGA hydrogel to promote vascular and hair follicle regeneration, therefore aiding wound healing, was assessed in a full-thickness skin defect model in mice. According to the findings, the PDGA hydrogel significantly improved wound healing and hemostasis in skin tissue defects. Because of the need for quick, safe, and efficient wound care techniques in the surgical settings, it enables quick control of bleeding events, which is especially crucial in hemostasis. With the aim of healing challenging skin wounds, electro-responsive click-hydrogels made of hyaluronic acid (clickHA) crosslinked with a modified polyethylene glycol precursor (PEG) were produced by oxidatively polymerizing a conducting polymer, poly(hydroxymethyl-3,4-ethylenedioxythiophene) (PEDOT-MeOH) [46]. These hydrogels exhibited significant biocompatibility and the conducting hydrogel formulation facilitates effective cell migration by electrostimulation with a low voltage (0.5 V) over a brief period of time (15 min). As a result,

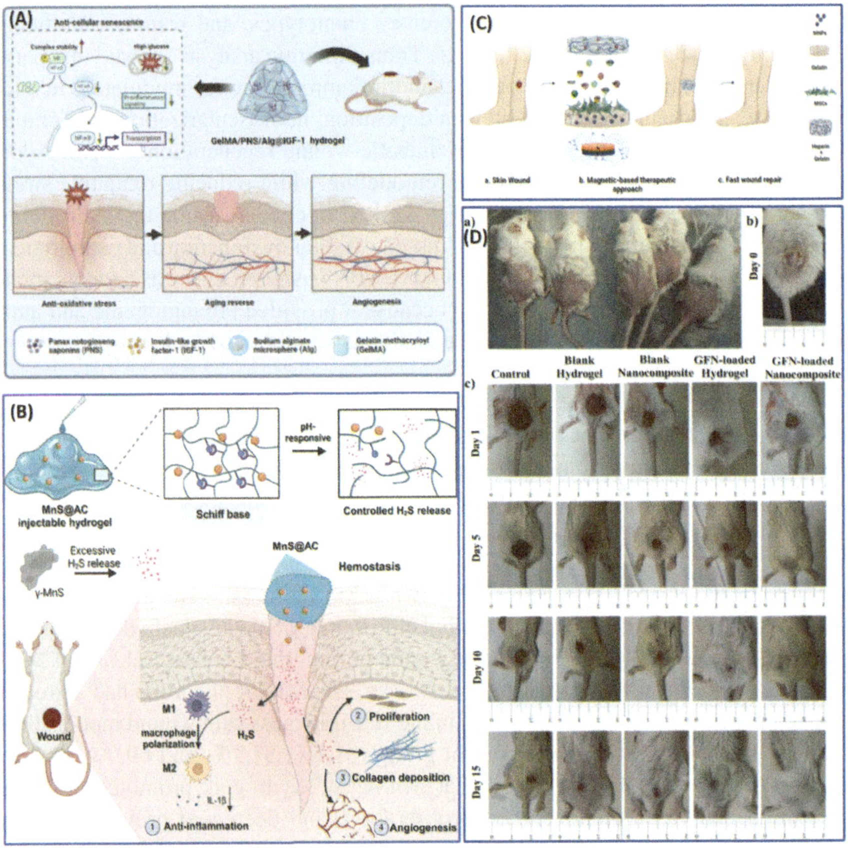

Fig. 6.4 **A** Mechanism of diabetic wound healing by GelMA/PNS/Alg@IGF-1 hydrogel. Reproduced with permission from Ref. [43]. Copyright 2025 Springer Nature (Open access). **B** Injectable polysaccharide hydrogels containing α-phase MnS NPs to create pH-responsive wound dressings (MnS@AC) with significant reduction of proinflammation and promote collagen deposition, angiogenesis, and proliferation. Reproduced with permission from Ref. [44]. Copyright 2025 Elsevier. **C** Wound repair using magnetic gelatin ferrogels by promoting angiogenesis to improve tissue regeneration in a variety of tissue damage conditions, such as skin wounds, bone abnormalities, and cardiovascular disorders. Reproduced with permission from Ref. [47]. Copyright 2024 American Chemical Society. **D** The preparation of mouse groups for the assessment of wound healing is shown schematically in (**a**); the gross appearance of the wound on day (0) is shown in (**b**); the appearance of the wounds without treatment and those treated with blank and GFN-loaded hydrogel and nanocomposite on days (1), (5), (10), and (15) are shown in (**c**). Reproduced with permission from Ref. [48]. Copyright 2020 Elsevier

the wound gap was closed in about one hour, and a uniform monolayer of migrating cells developed.

Ferrogels have shown significant potential in soft tissue regeneration, especially in wound healing. Their unique responsiveness to external magnetic fields enables controlled drug delivery, targeted cell stimulation, and enhanced tissue regeneration.

Ferrogels may be designed to precisely and locally release therapeutic chemicals, such as growth factors, antibiotics, or anti-inflammatory substances, minimizing systemic adverse effects and accelerating recovery. Furthermore, their porous and biocompatible nature promotes cell adhesion and proliferation, both of which are necessary for tissue healing. Wound healing may be accelerated by magnetic stimulation, which can also increase angiogenesis and collagen production. Ferrogels have the potential to cure chronic wounds like diabetic ulcers, where traditional treatments often fall short, according to recent research. Ferrogels are a promising, multipurpose platform that combines controlled treatment, responsive behaviour, and structural support to solve complicated wound healing difficulties and improve patient outcomes. Magnetic gelatin hydrogels were developed for this purpose in recent research, either with or without heparin surface modification, to improve angiogenesis and wound healing outcomes [47]. According to the findings of the study, (1) heparinizing gelatin scaffolds give the hydrogel a three-dimensional structure that mimics the extracellular matrix, increase its bioactivity, and promote the adhesion and proliferation of mesenchymal stem/stromal cells (MSCs); (2) different concentrations of heparin and magnetic nanoparticles in the hydrogel gave distinct biophysical cues to direct the behavior of MSCs seeded on them; and (3) heparinizing scaffolds and magnetic stimulation combined to promote cell proliferation and the synthesis of critical growth factors that promote healing.

A pH-sensitive magnetic nanocomposite hydrogel for the delivery of guaifenesin (GFN) for wound healing was created using itaconic acid onto starch when Fe_3O_4 nanoparticles were present via graft copolymerization [48]. Only around 54.1 GFN was delivered after one day for free Fe_3O_4-nanocomposite; however, that was 90.4% for the nanocomposite containing % 0.83 Fe_3O_4 under the same conditions. This is because the introduction of an external magnetic field may significantly increase the drug release percentage. More significantly, the bioavailability of the released GFN was maintained. The as-synthesized hydrogels' wound-healing properties were assessed using mice as a group model (Fig. 6.4D a). The wound healing process of defect regions is depicted in Fig. 6.4D b, c. It was shown that after 20 days, the wounds of the hydrogel and nanocomposite that were blank and filled with GFN had healed. However, the GFN-loaded nanocomposite's wound healed after ten days, and the nanocomposites' wound circles were noticeably smaller than the hydrogels'. This outcome might be explained by the strong adhesion of the composite hydrogels to mucosa and moist tissues, as well as the anti-inflammatory qualities of GFN produced to inhibit the growth of bacteria. These hydrogel drug carriers provide a suitable platform for magnetically targeted drug delivery as well as a dressing for wound healing.

6.5 HG-FGs in Imaging Diagnostics

Hydrogels and ferrogels have emerged as important materials in the realm of biomedical diagnostics, owing to their unique physicochemical properties and functional versatility. Recent advancements have expanded their applications across various diagnostic modalities, enhancing sensitivity, specificity, and real-time monitoring capabilities. There are several uses and research possibilities for ferrogel scaffolds. For instance, the difference in the two materials' acoustic impedances determines how much ultrasonic light is reflected at the ferrogel/water contact. It has previously been shown how magnetic nanoparticles affect the ultrasonic characteristics of gel phantoms, which also comprise acoustic impedance. Due to the large number of scatterers, the phantoms with magnetic nanoparticles showed improved echogenicity with the inclusion of nanostructures, as perceived by the advanced impedance values [49]. The mobility of ferrogel-based micro-objects impacted by an external magnetic field may often be controlled by ferrogels' high magnetic responses [50]. Ferrogel-based ultrasonography study was published [51], irrespective of the location or form of the sample. The amount of reflected echo signal increases with the amount of magnetic nanoparticles present in the gel. This finding is not directly related to how the magnetic nanoparticles affect the intensity of the reflected echo signal, since the ultrasonic effect that is employed has a wavelength that is far longer than the particle size. To understand the experimental results, a qualitative theoretical model was proposed, taking into account the hypothesis that the water in the gel's porous structure, the macromolecular net, and the hydrogel undergo viscous Stokes-like interactions as a result of the hydrogel's acoustic oscillations. Additionally, according to Tzitzios and colleagues [52], ferrogels are a viable option for magnetic resonance imaging (MRI) due to their non-contact characteristics and control input. The outcome showed that magnetic Fe_2O_3 nanoparticles embedded in the laponite matrix had a very good T2-weighted MRI contrast. The composite maghemite/laponite particles with a hydrodynamic diameter of 200 nm generate very stable gels, depending on their water content.

Thus, because of their combined therapeutic and diagnostic properties, HG-FGs, a multifunctional material, provide a number of benefits. First of all, HG-FGs provide effective drug delivery methods and real-time biomarker monitoring, facilitating timely disease detection and assessment. With this dual capability, medical practitioners may better understand how the illness develops throughout therapy, allowing them to adjust treatment plans and increase effectiveness. Secondly, the better formulation design flexibility of HG-FGs allows for the alteration of their physical and chemical properties, such as adhesion, rheology, and degradation rates, to provide more customized, superior treatments. This flexibility is essential for enhancing biocompatibility and controlling drug release inside the patient's body, which reduces side effects and increases patient comfort. Furthermore, HG-FGs often include naturally occurring biopolymers or modified synthetic polymers, which provide enhanced biodegradability and biocompatibility, enhancing safety and usefulness in clinical settings. HG-FGs are gradually heading toward future applications as nanotechnology advances. In reaction to environmental changes, such as pH, temperature, and biological signals, they are able to modify adaptive treatment approaches. This smart response significantly improves patient outcomes and quality of life by creating new opportunities in fields like cancer therapy and diabetes care. To sum up, HG-FGs exhibit a variety of possible uses in modern medicine. Precision and tailored medicine are made possible by their stimuli-responsive properties, better tunability, exceptional biocompatibility, and multifunctionality. The many biological uses of hydrogels and ferrogels are described in Table 6.1, with special attention paid to their distinct compositions, functions, and modes of action. Because of their increased water content and adjustable physical characteristics, hydrogels, which are usually made of biocompatible polymers like poly(ethylene glycol) (PEG), alginate, or chitosan, are used for tissue engineering, wound cure, and controlled drug delivery. Ferrogels, which incorporate magnetic nanoparticles like iron oxide into hydrogel matrices, are used for specific purposes, including remote-controlled actuation in implantable devices or soft robotics and targeted medication delivery. These materials respond to physiological or externally applied stimuli (e.g., pH, temperature, or magnetic fields), enabling precise control over therapeutic agent release or mechanical behaviour; consequently, adverse effects are decreased and treatment effectiveness is increased.

6.6 Summary and Prospects

To minimize the negative effects of ineffective treatments, theranostics integrates diagnosis and treatment, two comparatively independent approaches to disease management. Its goal is to promote individualized healthcare, based on molecular-level diagnosis, accurate disease information, personalized medication based on each patient's unique circumstances, and real-time patient response monitoring. Building such a bi-directional functional platform with HG-FGs is a potential approach. Combining diagnostic instruments with drugs or therapies to shorten the time between diagnosis and treatment and increase the overall effectiveness of patient care is an example of how effective HG-FGs are. Additionally, HG-FGs show improved mechanical properties, controlled release, and biocompatibility and may be customized to fulfill a variety of clinical needs. The characteristics of HG-FGs make them promising in the medical field. Despite the many advantages of HG-FGs, there are still certain problems and challenges with the current situation. To begin with, developing HG-FGs materials with different therapeutic and diagnostic qualities while ensuring their efficacy and compatibility with drugs and medical devices is a difficult technological undertaking. Second, the proper hydrogel or ferrogel materials must be selected in order for theranostic functions to be effective. These materials must have good biocompatibility, prolonged drug release, and reliable diagnostic properties. Finally, before HG-FGs may be used for therapeutic purposes in the real world, extensive clinical validation is required. This means evaluating factors including their efficacy, safety, and long-term effects. But as science and technology

Table 6.1 Biomedical applications of hydrogels and ferrogels, detailing the specific materials used, their functional purpose, and the underlying mechanisms enabling their therapeutic actions

S. No.	Application	Type of hydrogel/ferrogel	Key functions	Mechanism of application
1	Diabetes management	Poly(N-isopropylacrylamide) (PNIPAAm)/chitosan	Glucose-responsive insulin delivery	Swelling-induced release triggered by glucose level changes
2	Wound healing	Alginate/chitosan/gelatin	Moist healing environment, antimicrobial release	Forms a moist gel matrix, absorbs exudate, delivers antimicrobials
3	Bone tissue engineering	Fe_3O_4 nanoparticle-loaded gelatin methacrylate (GelMA)	Magnetic stimulation for osteogenesis and bone regeneration	Magnetic field enhances nanoparticle-mediated stimulation of bone cells
4	Tumor drug delivery	Magnetite (Fe_3O_4) + polyacrylamide	Magnetically triggered localized chemotherapy	External magnetic field triggers drug release at tumor site
5	Cartilage repair	Hyaluronic acid (HA)/PEGDA	Scaffold for chondrocyte growth and ECM production	Hydrogel supports cartilage cell growth and matrix deposition
6	Neural tissue engineering	Polypyrrole-gelatin/alginate	Supports neural cell attachment and conductivity	Conductive hydrogel transmits electrical signals, aiding neuron repair
7	Osteoarthritis treatment	Hyaluronic acid/chitosan	Injectable, joint lubrication, anti-inflammatory delivery	Viscous gel mimics synovial fluid, reduces inflammation
8	Tumor hyperthermia therapy	Fe_3O_4-PVA (polyvinyl alcohol)	Heat induction to kill cancer cells (magnetic hyperthermia)	Magnetic heating of embedded nanoparticles causes localized tumor cell death
9	Eye surgery (corneal repair)	Collagen/PEG-based hydrogel	Transparent, biocompatible scaffold for corneal tissue	Hydrogel integrates with corneal tissue, supporting cell growth
10	Pancreatic islet encapsulation	Alginate/PEG hydrogel	Immunoisolation of transplanted islets	Encapsulates islets in a semi-permeable membrane to block immune cells
11	Cardiac tissue regeneration	Fibrin/GelMA/collagen hydrogel	Supports myocardial regeneration post-infarction	Injectable gel delivers cells/drugs to infarct zone, supports cell adhesion

(continued)

Table 6.1 (continued)

S. No.	Application	Type of hydrogel/ferrogel	Key functions	Mechanism of application
12	Gene delivery systems	Chitosan-Fe_3O_4 nanocomposite	Magnetically guided gene delivery	Magnetic field guides gene-loaded ferrogel to target site
13	Spinal cord injury repair	Methylcellulose/ chitosan-collagen hydrogel	Axonal growth support, neuroregeneration	Supports axonal guidance and releases growth factors
14	3D bioprinting	GelMA/alginate/PEGDA bioinks	Printable scaffold with tunable bioactivity	Crosslinked network forms printable structures that mimic ECM
15	Biosensors (e.g., glucose)	Polyacrylamide + glucose oxidase/Fe_3O_4 composite	Responsive platform for real-time glucose sensing	Bioreactive response to analyte changes enables sensing function

progress and these problems continue to be solved, it is expected that HG-FGs will be manufactured and used more widely in the future.

This chapter summarizes HG-FGs' theranostic uses and identifies possible areas for development. Its potential for development in lab and clinical settings was investigated in order to make it a crucial part of biology, medicine, and other fields. HG-FGs have the ability to distribute drugs to treat diseases, respond to external stimuli by altering their structure, and detect sick signals or environmental changes. Despite recent improvements, the development of HG-FGs remains challenging until broad adoption in real-world applications. For example, the use of medicine poses a major challenge to scientists in developing complex and unique pathologic signal detection systems as well as accurate drug-controlled release systems. To further address practical problems in clinical practice, we should focus on developing new pharmaceutical materials, hydrogels, multidisciplinary collaboration, and information sharing. Furthermore, the creation of HG-FGs opens up new avenues for medical treatment; nevertheless, comprehensive studies and clinical trials are also required to confirm their safety and effectiveness before practical uses can be realized. In conclusion, although HG-FGs research is still at an early stage, these materials show significant potential and hold considerable advancement in the medical industry.

References

1. Y. Zheng, S.H. Ley, F.B. Hu, Global aetiology and epidemiology of type 2 diabetes mellitus and its complications. Nat. Rev. Endocrinol. **14**, 88–98 (2018). https://doi.org/10.1038/nrendo.2017.151
2. Z. Guan, H. Li, R. Liu, C. Cai, Y. Liu, J. Li, X. Wang, S. Huang, L. Wu, D. Liu, S. Yu, Z. Wang, J. Shu, X. Hou, X. Yang, W. Jia, B. Sheng, Artificial intelligence in diabetes management:

advancements, opportunities, and challenges. Cell Rep. Med. **4**, 101213 (2023). https://doi.org/10.1016/j.xcrm.2023.101213
3. R. Yin, K. Wang, S. Du, L. Chen, J. Nie, W. Zhang, Design of genipin-crosslinked microgels from concanavalin A and glucosyloxyethyl acrylated chitosan for glucose-responsive insulin delivery. Carbohydr. Polym. **103**, 369–376 (2014). https://doi.org/10.1016/j.carbpol.2013.12.067
4. A.K. Yetisen, N. Jiang, A. Fallahi, Y. Montelongo, G.U. Ruiz-Esparza, A. Tamayol, Y.S. Zhang, I. Mahmood, S. Yang, K.S. Kim, H. Butt, A. Khademhosseini, S. Yun, Glucose-sensitive hydrogel optical fibers functionalized with phenylboronic acid. Adv. Mater. **29** (2017). https://doi.org/10.1002/adma.201606380
5. Y. Liang, M. Li, Y. Yang, L. Qiao, H. Xu, B. Guo, PH/glucose dual responsive metformin release hydrogel dressings with adhesion and self-healing via dual-dynamic bonding for athletic diabetic foot wound healing. ACS Nano **16**, 3194–3207 (2022). https://doi.org/10.1021/acsnano.1c11040
6. N.A. Emad, J. Pandit, A. Ali, A. Rathee, P. Solanki, K. Imtiyaz, M.M.A. Rizvi, M. Aqil, M.A. Khan, Y. Sultana, Beeswax-based nanoconstructs enriched dual responsive hydrogel for diabetic foot ulcers in streptozotocin-induced diabetic rats. Int. J. Biol. Macromol. **288**, 138500 (2025). https://doi.org/10.1016/j.ijbiomac.2024.138500
7. S.S. Mandal, A.M. Choudhury, A. Gupta, P. Maiti, An injectable cyclodextrin extended polyurethane/carboxymethyl cellulose hydrogel for controlled release of insulin: in-vitro and in-vivo diabetic animal model study. Carbohydr. Polym. **356**, 123396 (2025). https://doi.org/10.1016/j.carbpol.2025.123396
8. X. Xing, Z. Gong, C. Chen, Y. Lin, P. Liu, T. Xiao, H. Yu, Y. Li, Y. Lin, G. Tan, C. Ning, Z. Wu, L. Wang, L. Zhou, Injectable bioresponsive bone adhesive hydrogels inhibit NLRP3 inflammasome on demand to accelerate diabetic fracture healing. Biomaterials **317**, 123059 (2025). https://doi.org/10.1016/j.biomaterials.2024.123059
9. R. Patwa, N. Saha, P. Sáha, Magnetic hydrogel based shoe insoles for prevention of diabetic foot. J. Magn. Magn. Mater. **514**, 167153 (2020). https://doi.org/10.1016/j.jmmm.2020.167153
10. J. Guo, F. Wang, Y. Hu, Y. Luo, Y. Wei, K. Xu, H. Zhang, H. Liu, L. Bo, S. Lv, S. Sheng, X. Zhuang, T. Zhang, C. Xu, X. Chen, J. Su, Exosome-based bone-targeting drug delivery alleviates impaired osteoblastic bone formation and bone loss in inflammatory bowel diseases. Cell Rep. Med. **4**, 100881 (2023). https://doi.org/10.1016/j.xcrm.2022.100881
11. Y. Tao, Y. Chen, S. Wang, W. Chen, D. Zhou, D. Chen, C. Zhang, Z. Wu, J. Yan, H. Zhang, Y. Wei, J. Su, Optimizing the modification density of acid oligopeptides to enhance the bone-targeting activity of liposomes. Compos. B Eng. **247**, 110288 (2022). https://doi.org/10.1016/j.compositesb.2022.110288
12. Y. Hu, X. Chen, S. Wang, Y. Jing, J. Su, Subchondral bone microenvironment in osteoarthritis and pain. Bone Res. **9**, 20 (2021). https://doi.org/10.1038/s41413-021-00147-z
13. A. Ceriello, K. Esposito, L. Piconi, M.A. Ihnat, J.E. Thorpe, R. Testa, M. Boemi, D. Giugliano, Oscillating glucose is more deleterious to endothelial function and oxidative stress than mean glucose in normal and type 2 diabetic patients. Diabetes **57**, 1349–1354 (2008). https://doi.org/10.2337/db08-0063
14. D. Li, K. Chen, H. Tang, S. Hu, L. Xin, X. Jing, Q. He, S. Wang, J. Song, L. Mei, R.D. Cannon, P. Ji, H. Wang, T. Chen, A logic-based diagnostic and therapeutic hydrogel with multistimuli responsiveness to orchestrate diabetic bone regeneration. Adv. Mater. **34** (2022). https://doi.org/10.1002/adma.202108430
15. S. Awasthi, S.K. Pandey, H.J. Shwetha, Nehal, S. Selvaraj, Tough, durable and strongly bonded self-healing cartilage-mimicking noncovalently assembled hydrogel nanostructures: the interplay between experiment and theory. Nanoscale **17**, 1616–1643 (2025). https://doi.org/10.1039/D4NR03322A
16. T. Naolou, N. Schadzek, J.M. Hornbostel, I. Pepelanova, M. Frommer, F. Lötz, L. Sauheitl, S. Dultz, V.J.M.N.L. Felde, O. Myklebost, C. Lee-Thedieck, Enhanced gelatin methacryloyl nanohydroxyapatite hydrogel for high-fidelity 3D printing of bone tissue engineering scaffolds. Biofabrication **17**, 025033 (2025). https://doi.org/10.1088/1758-5090/adbb90

17. A.C. Sousa, R. Alvites, B. Lopes, P. Sousa, A. Moreira, A. Coelho, A. Rêma, S. Biscaia, R. Cordeiro, F. Faria, G.F. da Silva, I. Amorim, J.D. Santos, L. Atayde, N. Alves, M. Domingos, A.C. Maurício, Hybrid scaffolds for bone tissue engineering: integration of composites and bioactive hydrogels loaded with hDPSCs. Biomater. Adv. **166**, 214042 (2025). https://doi.org/10.1016/j.bioadv.2024.214042
18. H. Deng, X. Chen, W. Zhang, H. Yang, X. Yang, Q. Zhang, Y. Yan, Bioactive glasses-based nanoenzymes composite double-network hydrogel with ROS scavenging for bone tissue engineering. Eur. Polym. J. **224**, 113718 (2025). https://doi.org/10.1016/j.eurpolymj.2025.113718
19. C. Gao, M. Feng, Y. Feng, P. Zhu, Q. Gao, Adhesive, injectable, and conductive poly(γ-glutamic acid) hydrogels reinforced by iron-doped mesoporous bioglass nanospheres for bone tissue engineering. J. Appl. Polym. Sci. (2025). https://doi.org/10.1002/app.56983
20. H. Liu, H. Zhang, Y. Han, Y. Hu, Z. Geng, J. Su, Bacterial extracellular vesicles-based therapeutic strategies for bone and soft tissue tumors therapy. Theranostics **12**, 6576–6594 (2022). https://doi.org/10.7150/thno.78034
21. A.C. Marques, P.J. Costa, S. Velho, M.H. Amaral, Stimuli-responsive hydrogels for intratumoral drug delivery. Drug Discov. Today **26**, 2397–2405 (2021). https://doi.org/10.1016/j.drudis.2021.04.012
22. S. Eom, S.G. Park, Y. Koo, Y. Noh, C. Choi, Y. Kim, H. Jun, C. Cha, J. Joo, S. Kang, In situ forming and self-crosslinkable protein hydrogels for localized cancer therapy and topical wound healing. J. Control. Release **378**, 460–475 (2025). https://doi.org/10.1016/j.jconrel.2024.12.026
23. J. Pennells, D.J. Martin, P.K. Annamalai, Nanocellulose: a sustainable nanomaterial for controlled drug delivery applications, in *Innovation in Nano-Polysaccharides for Eco-Sustainability* (Elsevier, 2022), pp. 217–253. https://doi.org/10.1016/B978-0-12-823439-6.00010-6
24. R. Foroutan, A. Mohammadzadeh, S. Javanbakht, R. Mohammadi, M. Ghorbani, Alginate/magnetic hydroxyapatite bio-nanocomposite hydrogel bead as a pH-responsive oral drug carrier for potential colon cancer therapy. Results Chem. **15**, 102177 (2025). https://doi.org/10.1016/j.rechem.2025.102177
25. X. Long, J. Wang, H. Wang, K. Hu, W. Zhang, W. Lin, C. Fang, K. Cheng, Z. Song, Injectable 2D-MoS_2-integrated bioadhesive hydrogel as photothermal-derived and drug-delivery implant for colorectal cancer therapy. Adv. Healthc. Mater. **14** (2025). https://doi.org/10.1002/adhm.202404842
26. M. Kim, J. Lee, J. Choi, S.Y. Kim, Reactive oxygen species-responsive nanocomposite hydrogels for accurate drug delivery and localized PDT/PTT/chemo synergistic cancer therapy. Eur. Polym. J. **224**, 113683 (2025). https://doi.org/10.1016/j.eurpolymj.2024.113683
27. Y. Xu, M. Gong, Y. Wang, Y. Yang, S. Liu, Q. Zeng, Global trends and forecasts of breast cancer incidence and deaths. Sci. Data **10**, 334 (2023). https://doi.org/10.1038/s41597-023-02253-5
28. S.-B. Kwak, S.J. Kim, J. Kim, Y.-L. Kang, C.W. Ko, I. Kim, J.-W. Park, Tumor regionalization after surgery: roles of the tumor microenvironment and neutrophil extracellular traps. Exp. Mol. Med. **54**, 720–729 (2022). https://doi.org/10.1038/s12276-022-00784-2
29. Z. Chen, P. Zhang, Y. Xu, J. Yan, Z. Liu, W.B. Lau, B. Lau, Y. Li, X. Zhao, Y. Wei, S. Zhou, Surgical stress and cancer progression: the twisted tango. Mol. Cancer **18**, 132 (2019). https://doi.org/10.1186/s12943-019-1058-3
30. M. Alieva, J. van Rheenen, M.L.D. Broekman, Potential impact of invasive surgical procedures on primary tumor growth and metastasis. Clin. Exp. Metastasis **35**, 319–331 (2018). https://doi.org/10.1007/s10585-018-9896-8
31. N. Vasan, J. Baselga, D.M. Hyman, A view on drug resistance in cancer. Nature **575**, 299–309 (2019). https://doi.org/10.1038/s41586-019-1730-1
32. D. Basak, S. Arrighi, Y. Darwiche, S. Deb, Comparison of anticancer drug toxicities: paradigm shift in adverse effect profile. Life **12**, 48 (2021). https://doi.org/10.3390/life12010048

33. R. Fan, Y. Cheng, R. Wang, T. Zhang, H. Zhang, J. Li, S. Song, A. Zheng, Thermosensitive hydrogels and advances in their application in disease therapy. Polymers (Basel) **14**, 2379 (2022). https://doi.org/10.3390/polym14122379
34. J. Liu, Controlled release of paclitaxel from a self-assembling peptide hydrogel formed in situ and antitumor study in vitro. Int. J. Nanomed. 2143 (2011). https://doi.org/10.2147/IJN.S24038
35. A.S. Mikhail, R. Morhard, M. Mauda-Havakuk, M. Kassin, A. Arrichiello, B.J. Wood, Hydrogel drug delivery systems for minimally invasive local immunotherapy of cancer. Adv. Drug Deliv. Rev. **202**, 115083 (2023). https://doi.org/10.1016/j.addr.2023.115083
36. H. Xin, S. Naficy, Drug delivery based on stimuli-responsive injectable hydrogels for breast cancer therapy: a review. Gels **8**, 45 (2022). https://doi.org/10.3390/gels8010045
37. M. Kumar, A. Jha, P. Goswami, R. Srivastava, M. Manjit, K. Bharti, B. Koch, B. Mishra, Fabrication of lipid-modified drug nanocrystals loaded injectable hydrogel for breast cancer therapy. Discov. Nano **20**, 30 (2025). https://doi.org/10.1186/s11671-025-04195-w
38. G. Verma, P. Gautam, S.B. Shelar, K.C. Barick, P.A. Hassan, Gelatin grafted magnetic hydroxyapatite nanocomposite for chemo-hyperthermia therapy. J. Magn. Magn. Mater. **625**, 173086 (2025). https://doi.org/10.1016/j.jmmm.2025.173086
39. N.S. Sumitha, N.G. Krishna, G.S. Sailaja, Multifunctional chitosan ferrogels for targeted cancer therapy by on-demand magnetically triggered drug delivery and hyperthermia. Biomater. Adv. **142**, 213137 (2022). https://doi.org/10.1016/j.bioadv.2022.213137
40. C. Kim, H. Kim, H. Park, K.Y. Lee, Controlling the porous structure of alginate ferrogel for anticancer drug delivery under magnetic stimulation. Carbohydr. Polym. **223**, 115045 (2019)
41. D. Lv, X. Cao, L. Zhong, Y. Dong, Z. Xu, Y. Rong, H. Xu, Z. Wang, H. Yang, R. Yin, M. Chen, C. Ke, Z. Hu, W. Deng, B. Tang, Targeting phenylpyruvate restrains excessive NLRP3 inflammasome activation and pathological inflammation in diabetic wound healing. Cell Rep. Med. **4**, 101129 (2023). https://doi.org/10.1016/j.xcrm.2023.101129
42. X. Sun, Y. Zhang, J. Cui, C. Zhang, C. Xing, H. Bian, J. Lv, D. Chen, L. Xiao, J. Su, Y. Liu, L. Su, Advanced multilayer composite dressing with co-delivery of gelsevirine and silk fibroin for burn wound healing. Compos. B Eng. **253**, 110549 (2023). https://doi.org/10.1016/j.compositesb.2023.110549
43. H. Yang, Y. Chen, Y. Rong, Y. Zhou, S. Li, X. Li, H. Wu, D. Lv, X. Cao, P. Wang, J. Zhu, B. Tang, Z. Hu, Multifunctional hydrogel targeting senescence to accelerate diabetic wound healing through promoting angiogenesis. J. Nanobiotechnol. **23**, 177 (2025). https://doi.org/10.1186/s12951-025-03274-5
44. L. Sun, Z. Zhang, Y. Cao, Z. Lu, Y. Zhang, X. He, Z. Wang, R. Pei, PH-responsive and on-demand hydrogen sulfide-releasing injectable hydrogel for wound healing promotion. Chem. Eng. J. **509**, 160904 (2025). https://doi.org/10.1016/j.cej.2025.160904
45. Y. Liu, Y. Zhang, Q. Jia, X. Liang, K. Xu, Rapid in situ formation of a double cross-linked network hydrogels for wound healing promotion. Front. Pharmacol. **16** (2025). https://doi.org/10.3389/fphar.2025.1562264
46. V. Castrejón-Comas, N. Mataró, L. Resina, D. Zanuy, Q. Nuñez-Aulina, J. Sánchez-Morán, H. Enshaei, M. Arnau, H. Muñoz-Galán, J.C. Worch, A.P. Dove, C. Alemán, M.M. Pérez-Madrigal, Electro-responsive hyaluronic acid-based click-hydrogels for wound healing. Carbohydr. Polym. **348**, 122941 (2025). https://doi.org/10.1016/j.carbpol.2024.122941
47. F. Pires, J.C. Silva, F.C. Ferreira, C.A.M. Portugal, Heparinized acellular hydrogels for magnetically induced wound healing applications. ACS Appl. Mater. Interfaces **16**, 9908–9924 (2024). https://doi.org/10.1021/acsami.3c18877
48. S. Nezami, M. Sadeghi, H. Mohajerani, A novel pH-sensitive and magnetic starch-based nanocomposite hydrogel as a controlled drug delivery system for wound healing. Polym. Degrad. Stab. **179**, 109255 (2020). https://doi.org/10.1016/j.polymdegradstab.2020.109255
49. K. Kaczmarek, T. Hornowski, B. Dobosz, A. Józefczak, Influence of magnetic nanoparticles on the focused ultrasound hyperthermia. Materials **11**, 1607 (2018). https://doi.org/10.3390/ma11091607
50. V.I. Zverev, A.P. Pyatakov, A.A. Shtil, A.M. Tishin, Novel applications of magnetic materials and technologies for medicine. J. Magn. Magn. Mater. **459**, 182–186 (2018). https://doi.org/10.1016/j.jmmm.2017.11.032

51. F.A. Blyakhman, S.Y. Sokolov, A.P. Safronov, O.A. Dinislamova, T.F. Shklyar, A.Y. Zubarev, G.V. Kurlyandskaya, Ferrogels ultrasonography for biomedical applications. Sensors **19**, 3959 (2019). https://doi.org/10.3390/s19183959
52. V. Tzitzios, G. Basina, A. Bakandritsos, C.G. Hadjipanayis, H. Mao, D. Niarchos, G.C. Hadjipanayis, J. Tucek, R. Zboril, Immobilization of magnetic iron oxide nanoparticles on laponite discs—an easy way to biocompatible ferrofluids and ferrogels. J. Mater. Chem. **20**, 5418 (2010). https://doi.org/10.1039/c0jm00061b

Chapter 7
Conclusions, Challenges, and Future Prospects of Hydrogels and Ferrogel

The Foundation of HG-FGs

Chapter 1 presented basic ideas for HG-FG, highlighting their smart behavior and operational structure. Hydrogels, with a hydrophilic nature, are considered soft, non-toxic with capability of self-healing. On the other hand, the addition of magnetic particles converts hydrogels to ferrogels, which can react under an external magnetic field. These HG-FG are now having multifunctional applications, including biomedical, cosmetics, environmental, etc. Chapter 1 developed a compact background for additional study into the strategy and functionality of HG-FGs.

Crosslinking, Design, and Physical Properties of Hg-FGs

The approaches related to crosslinking and design parameters, which are crucial for altering the mechanical, chemical, and biological properties of HG-FGs, were studied in Chap. 2. The various crosslinking methods, including chemical and physical crosslinking (covalent bonding, ionic bonding, and hydrogen bonding), contribute to the structural integrity and reactivity of soft gels. The chapter also give brief description of the conversion of hydrogels to ferrogels with a schematic mechanism. Polymerization techniques as well as physical characteristics with a special focus on the influence on swelling kinetics, mechanical, biological, and biodegradability of HG-FGs were also reviewed in Chap. 2. These discussions surely enhance the basic and advanced fundamentals of these soft materials.

Functionality of HG-FGs with Classifications

Chapter 3 classified hydrogels and ferrogels with respect to (i) origin, like natural vs. synthetic, (ii) preparation method, including monolithic, composite, and interpenetrating networks, (iii) structure, such as amorphous and crystalline, (iv) responsiveness, such as thermal, pH, magnetic, and photo-responsiveness. These basics of classifications of HG-FGs play an important role in understanding existing materials and leading the progress of these soft gels for future research. The hybrid materials are a combination of synthetic and natural gels and show improved structural,

S. Awasthi, *Smart Hydrogels and Ferrogels*, Materials Horizons: From Nature to Nanomaterials, https://doi.org/10.1007/978-981-95-9077-3_7

mechanical, tribological, and biological characteristics. The special focus was given to ferrogels, which show magneto-responsive properties that are crucial for triggered arrangements and controlled drug delivery systems.

Impact of Various Characterization Techniques on HG-FGs

Chapter 4 has conducted a comprehensive examination of the characterization techniques used to assess the properties and functioning of HG-FGs. A comprehensive analysis was performed on structural configuration, chemical composition, wear resistance, durability, and magnetic response of the gels. For this purpose, several techniques were utilized, such as scanning electron microscopy, x-ray diffraction, tribology, corrosion, and MTT assay. The chapter emphasized the need of many techniques to assess the potential of soft gels for diverse applications. The use of these approaches and their impact on HG-FGs provides a vital understanding of the dynamic behavior of HG-FGs in functional settings, hence improving application-specific tactics and predictive modelling.

Advanced Technologies: 3D, 4D, and Bioprinting

Chapter 5 gives an in-depth look at the latest developments in bioprinting, 3D, and 4D technologies that affect hydrogels and ferrogels. Laser-based methods, extrusion or nozzle-based methods, digital light processing, and inkjet printing are the key areas of interest in 3D printing. Every method is checked to see how well it prints, works with different materials, and can be used on a large scale in healthcare settings. This part looks at the switch to 4D printing, focusing on HG-FG systems and explaining how changes happen over time in response to inputs. The bioprinting part talks about how to combine tissues with printed parts. Topics include the advancement and refinement of bioinks for cellular encapsulation and proliferation, techniques to maintain cell viability and structural integrity during and post-printing, and practical applications in regenerative medicine, tissue engineering, and in vitro modelling. This chapter provides a comprehensive overview of the design and application tools for smart printed gels via the integration of additive manufacturing, bioengineering, and materials science.

Real-World Applications

Chapter 6 evaluated the uses of HG-FGs in translation across several domains, including diabetes, bone tissue engineering, cancer treatment, wound healing, and imaging diagnostics. The inquiry concentrated on its potential for advancement in laboratory and clinical settings to affirm its status as a fundamental element of biology, medicine, and other fields. Nonetheless, despite recent developments, the development of HG-FGs remains a viable material for translational applications. The use of medicine poses a considerable difficulty for researchers in creating complex and unique systems for detecting diseased signals and for accurate medication delivery mechanisms. To effectively solve problems in clinical practice, we need to focus on the development of new medicinal materials, hydrogels, working together across disciplines, and sharing information. Also, the progress made on HG-FGs opens up new medical treatment options. However, they need to be thoroughly tested in humans and animals to make sure they are safe and useful before they can be used. Finally, HG-FGs are a new area of study that is growing quickly, and these soft gels have gotten a lot of attention for their potential use in theranostics.

Challenges and Future Perspectives

However, noteworthy advancements have taken place; to effectively utilize the HG-FGs, there is still room which indeed requires alternative solutions. The scalability is one of the prime issues in translating such discoveries into industrial production due to its complexity and occasional expense; however, several reported research works based on synthesis tools facilitated evocative results on a small scale. Very importantly, in real-world applications, particularly in the field of biomedical and environmental conditions, it is essential to thoroughly verify the stability-related concerns. Moreover, one should carefully assess the smart materials like ferrogels allied with magnetic nanoparticles, which consist of long-standing environmental impact and biocompatibility. Notably and interestingly, in designing and predicting the nature of HG-FG, machine learning (ML) and artificial intelligence (AI) tools may be significantly helpful to accelerate the impactful and precise research outcomes for future developments. Furthermore, various suitable and best polymer compositions and their cross-linking approaches (including application-specific conformations) can be effectively discovered with the employment of good AI-skilled modelling tools as compared to some traditional trial-and-error-based tools. For using sustainable constituents and energy-efficient manufacturing processes, it will also be necessary to yield "green" HG-FGs in addressing the environmental issues and concerns.

Interdisciplinary Synergy: The Way Forward

Overall, this book marks significant evidence that multidisciplinary approaches are important for advances in HG-FGs. A cumulative network from materials scientists, chemists, and bioengineers is essential for real-world utilization of HG-FGs. The development of various technologies, such as nanotechnology, manufacturing processes, and biotechnology, must be balanced as per the legal framework and social demand. This type of equilibrium will justify technological practicability and perfect execution. Because of its flexible features, ability to respond to stimuli, and compatibility, the advanced research of HG-FGs has a big effect on healthcare, sustainable energy, and environmental technologies. The information in this book is a complete reference and a starting point for further research and growth. By fostering innovation and collaboration, we can optimally use HG-FGs to create a smarter, adaptive, and more sustainable environment.

The manufacturer's authorised representative in the EU is Springer Nature Customer Service Centre GmbH, Europaplatz 3, 69115 Heidelberg, Germany. If you have any concerns regarding our products, please contact ProductSafety@springernature.com

Printed and bound by CPI Group (UK) Ltd, Croydon, CR0 4YY
22/10/2026
02236452-0002